Springer Texts in Business and Economics

More information about this series at http://www.springer.com/series/10099

Victor A. Skormin

Introduction to Process Control

Analysis, Mathematical Modeling, Control and Optimization

 Springer

Victor A. Skormin
T.J. Watson School of Engineering
Binghamton University
Binghamton, NY, USA

ISSN 2192-4333   ISSN 2192-4341   (electronic)
Springer Texts in Business and Economics
ISBN 978-3-319-42257-2   ISBN 978-3-319-42258-9   (eBook)
DOI 10.1007/978-3-319-42258-9

Library of Congress Control Number: 2016947707

Printed on acid-free paper

This Springer imprint is published by Springer Nature
The registered company is Springer International Publishing AG
The registered company address is Gewerbestrasse 11, 6330 Cham, Switzerland

To Dr. Max Skormin, Isabel Skormin,
and (Little) Max Skormin

Foreword

This book is intended for graduate students and practitioners: it implies that readers have a serious background in at least one of the four areas. A professional statistician may discover how a regression model can facilitate control system design and/or provide basis for process optimization. A control system engineer may discover how to assess the accuracy of the traditional state-variable models. This book may help industrial leaders recognize optimization as an additional resource for improving process operation and motivate them to bring consultants. In the Internet era promoting the concepts of *recognizing the need for*, *knowing what is available*, and *understanding how it works* could be the most appealing feature of this book. The book is fully consistent with the author's motto, "I practice what I teach and teach what I practice."

Preface

Modern manufacturing facilities can be utilized to their full potential only when their operation is optimized. For the modern large-scale processes, even modest steps toward optimum result in very significant monetary gains. The optimization in Chap. 4 presents typical methods and problems of numerical optimization. These methods are readily available from various sources, but only to those who are aware of their existence and do recognize their applicability to particular situations. Once an optimal regime of operation is found, it has to be implemented and maintained over some time period. This task is hindered by adverse environmental effects and undesirable dynamic behavior of the process. Both conditions could be alleviated by (mostly discrete-time) control systems. Hence, Chap. 3 of this book features typical control problems and their practical solutions. Mathematical models provide a quantitative basis for the solution of optimization and control problems. In most instances process optimization cannot be implemented experimentally. In numerical optimization, mathematical models of the real processes are used as "guinea pigs" for testing various operational regimes. Control system design is based on mathematical models of the controlled processes in the form of transfer functions or matrix–vector descriptions. Therefore, Chap. 2 shows how mathematical models could be built and statistically validated. It includes cluster models that are not commonly known but are very useful in some situations. Some statistical and probabilistic concepts relevant to mathematical modeling are given in Chap. 1; however, some of these techniques offer useful tools for process analysis. Therefore, this book is not a book on optimization, control, mathematical modeling, or statistical analysis—numerous books on these subjects already exist—this book is intended to show how to apply these powerful disciplines as a set of tools to achieve a very important goal: improving the operation of modern manufacturing facilities (in chemistry, metallurgy, power generation, etc.)

This book reflects 40 years of the author's career as a university educator and industrial consultant, including professional experiences of his numerous former students. The author is grateful to Christina Stracquodaine and Matthew Davis for refining this text and specific numerical examples.

Binghamton, NY, USA   Victor A. Skormin

Contents

Introduction

Due to the complexity of manufacturing processes, process operation, control, and optimization constitute difficult tasks. Complexity of a typical manufacturing process implies incomplete knowledge of its physical nature, a large number of interrelated input and output process variables, noise in measurement channels, significant delays and inertia in the particular input–output channels, measurement inaccessibility of random factors relevant to process operation, and time dependence (drift) of process characteristics. Consequently, process operation is often based on the experience and intuition of the human operator and exhibits all problems labeled as "human errors." This approach does not allow utilization of existing manufacturing processes to their full potential. Numerically justified optimization and control methods can be implemented only on the basis of mathematical models and computers. This book will present mathematical modeling techniques and applications of model-based computer control and optimization methods intended for a practicing engineer.

Process Control

It is a quantitative world
We bend the properties of matter
We are invincible and bold
In going after bread and butter

We learn in the eternal school
We have an unforgiving teacher
We use a computer as a tool
Discover feature after feature

We use a statistical technique
In space of numerous dimensions
Reveal relations, strong and weak
For the sake of our intentions

We are on an unbiased way
With all the knowledge, strength, and vigor
And shaping mother nature's clay
Into magnificence of figure

We master rigors of control
Its theory and application
And never use a crystal ball
For our efforts' validation

We want to minimize the loss
Of time and money being spent
And utilizing the resource
To fullest possible extent

Chapter 1
Statistical Methods and Their Applications

In many ways, manufacturing processes are dominated by random phenomena that have to be characterized in order to describe the processes quantitatively. Random phenomena are the manifestation of "forces of nature," measurement errors, noise in the information channels, etc. Any attempt to characterize random phenomena on a case-by-case basis is meaningless: random phenomena manifest themselves differently under the same conditions. However, when a sufficiently large number of realizations, also known as a population of occurrences, of these phenomena are observed, one can detect general tendencies pertaining to the entire population. These general tendencies are the only way to characterize random phenomena. Statistics is a science that extracts general tendencies present in a large population of random events. In addition, statistics offers the means for assessing the dependability, or the limits of applicability, of the detected trends.

Random phenomena include *random events*, *random variables*, and *random processes*.

1.1 Random Events

A random event is an event that may or may not occur as the result of a trial. Random event is a concept that reflects the qualitative side of random phenomena. Typically, random events are labeled as A, B, C,... and are characterized by *probability*. Probability of a random event A, $P[A]$, is a positive number that does not exceed 1, i.e. $0 \leq P[A] \leq 1$. Event A is *impossible* if $P[A] = 0$; event A is *certain* if $P[A] = 1$.

Frequency of a random event A represents the likelihood of the occurrence of this event as the result of a trial. Unsurprisingly it is defined as the ratio $\frac{N^A}{N}$ where N is the total number of trials, and N^A is the number of trials where event A has occurred.

© Springer International Publishing Switzerland 2016
V.A. Skormin, *Introduction to Process Control*, Springer Texts
in Business and Economics, DOI 10.1007/978-3-319-42258-9_1

Probability is perceived as the limit of frequency as the total number of trials approaches infinity, $P[A] = \underset{N\to\infty}{Lim} \frac{N^A}{N}$.

Random events A and B constitute a *complete group of mutually exclusive events* if $P[A] + P[B] = 1$. This implies that any trial will result in the occurrence of only one of these two events.

In the situation when two events, A and B, can occur as the result of an experiment, the *conditional frequency* of event A subject to event B can be defined. It represents the likelihood of the occurrence of event A as the result of a trial providing that event B has already occurred in this trial. This is defined as the ratio $\frac{N^{AB}}{N^B}$, where N^B is the total number of trials that resulted in the occurrence of event B, and N^{AB} is the number of trials that resulted in the occurrence of both events, A and B. *Conditional probability* of event A subject to event B is viewed as the limit of conditional frequency as the total number of trials approaches infinity, i.e. $P[A/B] = \underset{N\to\infty}{Lim} \frac{N^{AB}}{N^B}$. Similarly, conditional probability of event B subject to event A is defined as $P[B/A] = \underset{N\to\infty}{Lim} \frac{N^{AB}}{N^A}$. Note that $N \to \infty$ inevitably leads to $N^A \to \infty$, $N^B \to \infty$, and $N^{AB} \to \infty$, however $N^{AB} \leq N^A$ and $N^{AB} \leq N^B$.

Event C, defined as a simultaneous occurrence of event A and event B, is called the product of two events, i.e. $C = A \cdot B$. It can be seen that the probability of event C is equal to the conditional probability of event A subject to event B times the probability of event B, i.e. $P[C] = P[A/B]P[B]$. Indeed, $P[A/B]P[B] = \underset{N\to\infty}{Lim} \frac{N^{AB}}{N^B} \underset{N\to\infty}{Lim} \frac{N^B}{N} = \underset{N\to\infty}{Lim} \left(\frac{N^{AB}}{N^B} \frac{N^B}{N} \right) = \underset{N\to\infty}{Lim} \frac{N^{AB}}{N} = P[AB] = P[C]$. Similarly, $P[C] = P[B/A]P[A]$ and therefore $P[A/B]P[B] = P[B/A]P[A]$.

Events A and B are called *independent events* if $P[A/B] = P[A]$.

Multiplication of Probabilities If A and B are independent events, $P[A/B] = P[A]$, then the probability of the simultaneous occurrence of two independent events, A and B, is equal to the product of the probabilities of each event, $P[AB] = P[A] \cdot P[B]$.

Addition of Probabilities If events A and B belong to a complete group of independent events, i.e. $P[A] + P[B] + P[C] + P[D] + \ldots = 1$, then the probability of occurrence of event A or event B as the result of a trial is defined as $P[A \text{ or } B] = P[A] + P[B]$, and the occurrence of A or B is called the sum of two events.

Bayes' Theorem If events A_1, A_2, and A_3 constitute a complete group of mutually exclusive events, i.e. $P[A_1] + P[A_2] + P[A_3] = 1$, and event B is related to A_1, A_2, and A_3 via conditional probabilities $P[B/A_1]$, $P[B/A_2]$, $P[B/A_3]$, then

$$P[B] = P[B/A_1] \cdot P[A_1] + P[B/A_2] \cdot P[A_2] + P[B/A_3] \cdot P[A_3].$$

Furthermore, if the number of trials N is very large then

$$P[B/A_1] \cdot P[A_1] + P[B/A_2] \cdot P[A_2] + P[B/A_3] \cdot P[A_3] \approx \frac{N^{BA_1}}{N^{A_1}} \cdot \frac{N^{A_1}}{N} + \frac{N^{BA_2}}{N^{A_2}} \cdot \frac{N^{A_2}}{N}$$

$$+ \frac{N^{BA_3}}{N^{A_3}} \cdot \frac{N^{A_3}}{N} = \frac{N^{BA_1} + N^{BA_2} + N^{BA_3}}{N}$$

where notations are self-explanatory.

Since A_1 or A_2 or A_3 always occurs, $N^{BA_1} + N^{BA_2} + N^{BA_3} = N^B$, $P[B/A_1] \cdot$
$P[A_1] + P[B/A_2] \cdot P[A_2] + P[B/A_3] \cdot P[A_3] \approx \frac{N^B}{N} \approx P[B]$

It is known that

$$P[B/A_1] \cdot P[A_1] = P[A_1/B] \cdot P[B] \text{ and}$$
$$P[B] = P[B/A_1] \cdot P[A_1] + P[B/A_2] \cdot P[A_2] + P[B/A_3] \cdot P[A_3]$$

This results in a very useful expression known as the Bayes' formula:

$$P[A_1/B] = \frac{P[B/A_1] \cdot P[A_1]}{P[B/A_1] \cdot P[A_1] + P[B/A_2] \cdot P[A_2] + P[B/A_3] \cdot P[A_3]}$$

Example 1.1 Application to quality prediction of the product at a glass manufacturing plant. The database contains 5500 data entries. This includes 879 cases of poor quality of the product. These 879 cases were exhibited during the following events:
- 136 cases of short-time electrical failure
- 177 cases of poor quality of the raw material
- 83 cases of minor equipment malfunction

The good quality cases include
- 36 cases of short-time electrical failure
- 81 cases of poor quality of the raw material
- 63 cases of minor equipment malfunction

Define the probability of having poor quality of the product if at 10 am a short-time electrical failure was reported, and later, at 2 pm poor quality of the raw material was observed.

Solution Assume that event A_1 — poor quality, event A_2 —good quality, note that $P[A_1] + P[A_2] = 1$, event B—occurrence of a short-time electrical failure, and event C — occurrence of poor quality of the raw material.

Based on the available data, the probabilities of getting poor quality product and good quality product are, $P[A_1] = 879/5500 = 0.16$ and $P[A_2] = (5500-879)/5500 = 0.84$, respectively. The conditional probability relating an electric failure to poor product quality is $P[B/A_1] = \frac{N^{BA_1}}{N_{A_1}} = \frac{136}{879} = .155$. The conditional probability relating an electric failure to good product quality is $P[B/A_2] = \frac{N^{BA_2}}{N_{A_2}} = \frac{36}{5500-879} = .008$.

Now, one can compute the probability of poor quality product subject to the short-time electrical failure reported at 10 am:

$$P[A_1/B] = \frac{P[B/A_1] \cdot P[A_1]}{P[B/A_1] \cdot P[A_1] + P[B/A_2] \cdot P[A_2]} = \frac{.155 \times .16}{.155 \times .16 + .008 \times .84} = .771$$

Example 1.2 Identification of the manufacturer of the catalytic converter.

There are 3 possible manufacturers of these converters. Catalytic converters are used to assure that the factory exhaust gases comply with the EPA requirements. They are expected to maintain their catalytic properties during 2000 h of operation with probabilities 0.83 (manufacturer #1), 0.87 (manufacturer #2), and 0.92 (manufacturer #3). It is known that all catalytic converters installed at the factory were produced by the same manufacturer. One of the converters failed after 1500 h of operation and the second failed after 1800 h. The manufacturer of the converters is unknown. What is the probability that manufacturer #3 produced them?

Solution Introduce events M_1, M_2, and M_3 representing appropriate manufacturers; it can be shown that since the manufacturer is unknown, then $P[M_1] = P[M_2] = P[M_3] = 0.3333$, therefore, $P[M_1] + P[M_2] + P[M_3] = 1$. Introduce event A to represent the failure of a catalyst during the first 2000 h of its operation, then

$$P[A/M_1] = 1 - 0.83 = 0.17$$
$$P[A/M_2] = 1 - 0.87 = 0.13$$
$$P[A/M_3] = 1 - 0.92 = 0.08$$

Now re-estimate the probability of M_1, M_2, and M_3 subject to the failure of the first converter:

$$P[M_1/A] = \frac{P[A/M_1] \cdot P[M_1]}{P[A/M_1] \cdot P[M_1] + P[A/M_2] \cdot P[M_2] + P[A/M_3] \cdot P[M_3]}$$
$$= \frac{.17 \times .3333}{.17 \times .3333 + .13 \times .3333 + .08 \times .3333} = .447$$
$$P[M_2/A] = \frac{.13 \times .3333}{.17 \times .3333 + .13 \times .3333 + .08 \times .3333} = .342$$
$$P[M_2/A] = \frac{.08 \times .3333}{.17 \times .3333 + .13 \times .3333 + .08 \times .3333} = .211$$

Re-estimate the probabilities of M_1, M_2, and M_3 subject to the failure of the second converter:

$$P[M_1/A] = \frac{.17 \times .447}{.17 \times .447 + .13 \times .342 + .08 \times .211} = .555$$

$$P[M_2/A] = \frac{.13 \times .342}{.17 \times .447 + .13 \times .342 + .08 \times .211} = .321$$

$$P[M_3/A] = \frac{.08 \times .211}{.17 \times .447 + .13 \times .342 + .08 \times .211} = .124,$$

and this is the answer.

It is good to realize that random events M_1, M_2, and M_3 constitute a full group of mutually exclusive events and the sum of their probabilities is equal to 1.0 after each reevaluation of their probabilities.

Example 1.3 Assessment of a student's expected success based on the previous performance.

According to departmental records out of 1200 students who took Automatic Control, 225 got grade A, 511 — grade B, 406 — grade C, 32 — grade D, and 26 — grade F. It is also known that among these groups the number of students who received grade B in Signals and Systems are correspondingly 67, 211, 108, 4, and 3. It is known that Sam J. just got a grade of B in Signals and Systems (event B^{SS}). What are his chances of getting grade A, grade B, and grade C in Controls?

Solution First, evaluate initial probabilities of getting various grades in Control:

$$P[A] = \frac{225}{1200} = .187, \quad P[B] = \frac{511}{1200} = .426, \quad P[C] = \frac{406}{1200} = .338,$$

$$P[D] = \frac{32}{1200} = .027, \quad P[F] = \frac{26}{1200} = .022$$

Now evaluate the following conditional probabilities:

$$P[B^{SS}/A] = \frac{67}{225} = .298, \quad P[B^{SS}/B] = \frac{211}{511} = .413, \quad P[B^{SS}/C] = \frac{108}{406} = .266,$$

$$P[B^{SS}/D] = \frac{4}{32} = .125, \quad P[B^{SS}/F] = \frac{3}{26} = .115$$

The resultant probabilities are as follows:

$$P[A/B^{SS}] =$$

$$\frac{P[B^{SS}/A] \times P[A]}{P[B^{SS}/A] \times P[A] + P[B^{SS}/B] \times P[B] + P[B^{SS}/C] \times P[C] + P[B^{SS}/D] \times P[D] + P[B^{SS}/F] \times P[F]}$$

$$= \frac{.298 \times .187}{.298 \times .187 + .413 \times .426 + .266 \times .338 + .125 \times .027 + .115 \times .022}$$

$$= \frac{.0557}{.0557 + .1759 + .0899 + .0034 + .0025} = .17$$

$$P[B/B^{SS}] =$$

$$\frac{P[B^{SS}/B] \times P[B]}{P[B^{SS}/A] \times P[A] + P[B^{SS}/B] \times P[B] + P[B^{SS}/C] \times P[C] + P[B^{SS}/D] \times P[D] + P[B^{SS}/F] \times P[F]}$$

$$= \frac{.413 \times .426}{.298 \times .187 + .413 \times .426 + .266 \times .338 + .125 \times .027 + .115 \times .022}$$

$$= \frac{.1759}{.0557 + .1759 + .0899 + .0034 + .0025} = .537$$

$$P[C/B^{SS}]$$

$$= \frac{P[B^{SS}/C] \times P[C]}{P[B^{SS}/A] \times P[A] + P[B^{SS}/B] \times P[B] + P[B^{SS}/C] \times P[C] + P[B^{SS}/D] \times P[D] + P[B^{SS}/F] \times P[F]}$$

$$= \frac{.266 \times .338}{.298 \times .187 + .413 \times .426 + .266 \times .338 + .125 \times .027 + .115 \times .022}$$

$$= \frac{.0899}{.0557 + .1759 + .0899 + .0034 + .0025} = .274$$

1.2 Random Variables

A random variable can have any numerical value, $x(k)$, at each trial, $k = 1,2,\ldots$ is the trial (realization) index. How can the general properties of a random variable be extracted? Consider an array: $x(k)$, $k = 1,2,\ldots,N$.

Detect minimum and maximum values of the variable within the array, X_{MIN} and X_{MAX}. Divide interval $[X_{MIN}, X_{MAX}]$ into M divisions

Define step $D_x = (X_{MAX} - X_{MIN})/M$.

Compute the number of realizations within each interval, i.e. number of realizations n_j such that $X_{MIN} + (j-1) \cdot D_x < X(k) \leq X_{MIN} + j \cdot D_x$, $j = 1,2,\ldots,M$

Compute frequencies $f_j = n_j/N$, $j = 1,2,\ldots,M$

Build a histogram showing frequencies f_j vs $x(k)$ values like the one shown below in Fig. 1.1 (it is said that a histogram relates values of a random variable to the frequency of occurrence of these values).

Then assume that $N \to \infty$, $M \to \infty$, and $D_x \to 0$. It can be seen that the histogram turns into a continuous line that represents the distribution law of the random variable $x(k)$, this is called the probability density $P(x)$.

The probability density function can be used to define the probability of random variable $x(k)$ which satisfies the condition $x_1 < x(k) \leq x_2$ as follows:

$$P[x_1 < x(k) \leq x_2] = \int_{x_1}^{x_2} P(x)dx = \int_{0}^{x_2} P(x)dx - \int_{0}^{x_1} P(x)dx$$

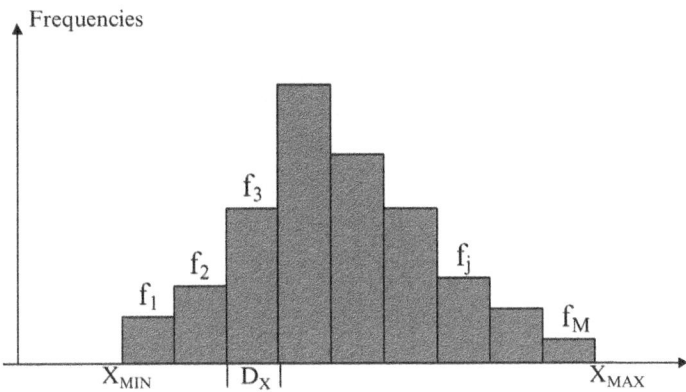

Fig. 1.1 Probability distribution function

One can realize that $P[-\infty < x(k) \le \infty] = \int\limits_{0}^{\infty} P(x)dx - \int\limits_{0}^{-\infty} P(x)dx = \int\limits_{0}^{\infty} P(x)dx +$

$\int\limits_{-\infty}^{0} P(x)dx = 1.$

It is known that there are several "typical" distribution laws, however, the most common is called *normal distribution*, which provides the best description of most random phenomena that can be observed in real life.

Normal Distribution has the following probability density:

$$P(x) = \frac{1}{\sigma\sqrt{2\pi}}\exp\left\{-\frac{(x-\mu)^2}{2\sigma^2}\right\} = P(x,\mu,\sigma),$$

where μ and σ are parameters of the normal distribution law: μ — mean value and σ — standard deviation. It is said that the normal distribution law reflects fundamental properties of nature. In terms of manufacturing, the standard deviation of any variable characterizing the product represents effects of "forces of nature" on a manufacturing process, and the mean value represents effects of operators' efforts and adjustment of equipment.

The following is the definition of the probability of a normally distributed random variable $x(k)$, satisfying the condition $x_1 < x(k) \le x_2$:

$$P[x_1 < x(k) \le x_2] = \int\limits_{x_1}^{x_2} P(x,\mu,\sigma)dx = \int\limits_{0}^{x_2} P(x,\mu,\sigma)dx - \int\limits_{0}^{x_1} P(x,\mu,\sigma)dx = \Phi(z_2) - \Phi(z_1)$$

where

$$z = \frac{x - \mu}{\sigma}, \ z_1 = \frac{x_1 - \mu}{\sigma}, \quad z_2 = \frac{x_2 - \mu}{\sigma}, \quad \Phi(z_1) = \frac{1}{\sqrt{2\pi}} \int_0^{z_1} \exp\left\{-\frac{z^2}{2}\right\} dz \quad \text{and}$$

$$\Phi(z_2) = \frac{1}{\sqrt{2\pi}} \int_0^{z_2} \exp\left\{-\frac{z^2}{2}\right\} dz.$$

Note that function $\Phi(z)$ is an odd function, i.e. $\Phi(-z) = -\Phi(z)$. This function is tabulated and is readily available. One should realize that this function continuously increases between zero and infinity, as shown below.

$$\Phi(0) = \frac{1}{\sqrt{2\pi}} \int_0^0 \exp\left\{-\frac{z^2}{2}\right\} dz = 0 \text{ and } \Phi(\infty) = \frac{1}{\sqrt{2\pi}} \int_0^\infty \exp\left\{-\frac{z^2}{2}\right\} dz = .5$$

Example 1.4 Evaluation of the effect of the improved automation

It is required to manufacture 300,000 special bolts. The factory's cost is $0.43 per unit. The allowable length of a bolt is between .194 and .204 inches. When the existing equipment is used, the length of manufactured bolts has a standard deviation of .003 inches. The introduction of an advanced control results in the reduction of this standard deviation to .00133 inches. Modification of the controls costs $5000. Determine if this modification would pay for itself.

Solution Since the length of bolts varies and only the "good" bolts will be accepted, the total number of manufactured bolts, N, will be greater than 300,000. Let us determine this number. It is expected that the equipment will be adjusted to assure the mean value of the length, $\mu = (x_1 + x_2)/2 = (.194 + .204)/2 = .199$ (inches). Now, when $\mu = .199$ and $\sigma = .003$ one can determine the probability of the length of a bolt to be within the allowable limits, $P[x_1 < x \le x_2]$, assuming the normal distribution law:

$$z_1 = (x_1 - \mu)/\sigma = (.194 - .199)/.003 = -1.67 \text{ and}$$
$$z_2 = (x_2 - \mu)/\sigma = (.204 - .199)/.003 = 1.67$$

Therefore, $P[x_1 < x \le x_2] = 2 \times \Phi(1.67)$. According to the table below in Fig. 1.2, $P[x_1 < x \le x_2] = 2 \times .4525 = .905$. This result indicates that in order to manufacture 300,000 "good" bolts, a total of $300,000/.905 = 331,492$ units must be produced.

Now let us repeat this calculation assuming the improved accuracy of the equipment (or reduced standard deviation, $\sigma = .00133$):

Standard Normal Distribution Table

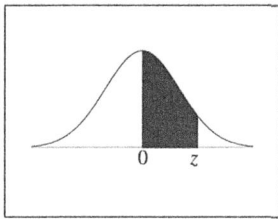

z	.00	.01	.02	.03	.04	.05	.06	.07	.08	.09
0.0	.0000	.0040	.0080	.0120	.0160	.0199	.0239	.0279	.0319	.0359
0.1	.0398	.0438	.0478	.0517	.0557	.0596	.0636	.0675	.0714	.0753
0.2	.0793	.0832	.0871	.0910	.0948	.0987	.1026	.1064	.1103	.1141
0.3	.1179	.1217	.1255	.1293	.1331	.1368	.1406	.1443	.1480	.1517
0.4	.1554	.1591	.1628	.1664	.1700	.1736	.1772	.1808	.1844	.1879
0.5	.1915	.1950	.1985	.2019	.2054	.2088	.2123	.2157	.2190	.2224
0.6	.2257	.2291	.2324	.2357	.2389	.2422	.2454	.2486	.2517	.2549
0.7	.2580	.2611	.2642	.2673	.2704	.2734	.2764	.2794	.2823	.2852
0.8	.2881	.2910	.2939	.2967	.2995	.3023	.3051	.3078	.3106	.3133
0.9	.3159	.3186	.3212	.3238	.3264	.3289	.3315	.3340	.3365	.3389
1.0	.3413	.3438	.3461	.3485	.3508	.3531	.3554	.3577	.3599	.3621
1.1	.3643	.3665	.3686	.3708	.3729	.3749	.3770	.3790	.3810	.3830
1.2	.3849	.3869	.3888	.3907	.3925	.3944	.3962	.3980	.3997	.4015
1.3	.4032	.4049	.4066	.4082	.4099	.4115	.4131	.4147	.4162	.4177
1.4	.4192	.4207	.4222	.4236	.4251	.4265	.4279	.4292	.4306	.4319
1.5	.4332	.4345	.4357	.4370	.4382	.4394	.4406	.4418	.4429	.4441
1.6	.4452	.4463	.4474	.4484	.4495	.4505	.4515	.4525	.4535	.4545
1.7	.4554	.4564	.4573	.4582	.4591	.4599	.4608	.4616	.4625	.4633
1.8	.4641	.4649	.4656	.4664	.4671	.4678	.4686	.4693	.4699	.4706
1.9	.4713	.4719	.4726	.4732	.4738	.4744	.4750	.4756	.4761	.4767
2.0	.4772	.4778	.4783	.4788	.4793	.4798	.4803	.4808	.4812	.4817
2.1	.4821	.4826	.4830	.4834	.4838	.4842	.4846	.4850	.4854	.4857
2.2	.4861	.4864	.4868	.4871	.4875	.4878	.4881	.4884	.4887	.4890
2.3	.4893	.4896	.4898	.4901	.4904	.4906	.4909	.4911	.4913	.4916
2.4	.4918	.4920	.4922	.4925	.4927	.4929	.4931	.4932	.4934	.4936
2.5	.4938	.4940	.4941	.4943	.4945	.4946	.4948	.4949	.4951	.4952
2.6	.4953	.4955	.4956	.4957	.4959	.4960	.4961	.4962	.4963	.4964
2.7	.4965	.4966	.4967	.4968	.4969	.4970	.4971	.4972	.4973	.4974
2.8	.4974	.4975	.4976	.4977	.4977	.4978	.4979	.4979	.4980	.4981
2.9	.4981	.4982	.4982	.4983	.4984	.4984	.4985	.4985	.4986	.4986
3.0	.4987	.4987	.4987	.4988	.4988	.4989	.4989	.4989	.4990	.4990
3.1	.4990	.4991	.4991	.4991	.4992	.4992	.4992	.4992	.4993	.4993
3.2	.4993	.4993	.4994	.4994	.4994	.4994	.4994	.4995	.4995	.4995
3.3	.4995	.4995	.4995	.4996	.4996	.4996	.4996	.4996	.4996	.4997
3.4	.4997	.4997	.4997	.4997	.4997	.4997	.4997	.4997	.4997	.4998
3.5	.4998	.4998	.4998	.4998	.4998	.4998	.4998	.4998	.4998	.4998

Fig. 1.2 Standard normal distribution table. http://unimasr.net/community/viewtopic.php?f=1791&t=82994

$z_1 = (.194-.199)/.00133 = -3.76$ and $z_2 = 3.76$, and $\Phi(3.76) \approx .4998$, therefore, $P[x_1 < x \le x_2] = .9996$. This indicates that the total of $300,000/.9996 = 300,121$ units must be produced, thus the effective savings of automation is $\$.43 \times (331,492 - 300,121) = \$13,489.53$.

The conclusion is obvious: the modification of controls is well justified.

Estimation of Mean and Variance The mean value of a random variable is estimated as $M_x = \dfrac{1}{N} \displaystyle\sum_{k=1}^{N} x(k)$. In some instances it is often said that $M_x = M_x(N)$ to emphasize that M_x is dependent on the number of realizations of the random variable that were used for the estimation. It is known that as $N \to \infty$ $M_x(N) \to \mu$, where μ is the appropriate parameter of the distribution law.

The variance and standard deviation of a random variable are estimated as

$V_x = \dfrac{1}{N-1} \displaystyle\sum_{k=1}^{N} [x(k) - M_x]^2$, and $S_x = \sqrt{V_x}$. Again, it is often said that $V_x = V_x(N)$ and $S_x = S_x(N)$ to emphasize that these estimates are dependent on the number of realizations of the random variable used for the estimation. It is known that as $N \to \infty$ $S_x(N) \to \sigma$, where σ is the appropriate parameter of the distribution law.

Recursive Estimation is common in the situations when characteristics of a random variable are calculated on-line. It is done to incorporate as many realizations, $x(k)$, as possible in the estimation without the penalty of storing an ever-increasing data array. The following formulae are applied:

$$M_x[N] = M_x[N-1] + \frac{1}{N}[x(N) - M_x[N-1]]$$

$$V_x[N] = V_x[N-1] + \frac{1}{N-1}[[x(N) - M_x(N)]^2 - V_x[N-1]], \text{ and } S_x[N] = \sqrt{V_x[N]}$$

The above expressions could be easily derived. Indeed,

$$M_x[N] = \frac{1}{N}\sum_{k=1}^{N} x(k) = \frac{1}{N}\left[x(N) + \sum_{k=1}^{N-1} x(k)\right] = \frac{1}{N}x(N) + \frac{N-1}{N(N-1)}\sum_{k=1}^{N-1} x(k)$$

$$= \frac{1}{N}x(N) + \frac{N-1}{N}M_x(N-1)$$

Finally, $M_x[N] = M_x[N-1] + \dfrac{1}{N}[x(N) - M_x(N-1)]$

Similar derivations could result in the formula for the estimation of variance.

Confidence Intervals for Mean Values and Variances One should realize that as with any statistical estimate, estimates of mean value and standard deviation extracted from N observations of random variable $x(k)$ are different from corresponding parameters of the distribution law. How is the discrepancy between the "true" and the estimated characteristics estimated? It is known that the estimation error is a random variable with its own statistical distribution. It is also known that estimation according to the above formulae results in *unbiased* estimates, i.e. the mean value of the estimation error is zero. The knowledge of the distribution law results in the following expression for the confidence interval for the mean value:

$$P\{M_x(N) - \Delta^M(N, \alpha) \leq \mu \leq M_x(N) + \Delta^M(N, \alpha)\} = 1 - 2\alpha$$

where $\Delta^M(N, \alpha) = t(N, \alpha) \dfrac{S_x(N)}{\sqrt{N}}$

The notations are,

$M_x(N)$ and $S_x(N)$ are mean value and standard deviation estimated using N observations,

$\mu = M_x(\infty)$ is the "true" mean value,

$\Delta^M(N, \alpha)$ is the width of the confidence interval that depends on the number of observations N and the significance level α,

$t(N, \alpha)$ is the t-distribution or Student distribution (named after the statistician who viewed himself as an eternal student). The t-distribution is tabulated and defined as a function of significance level α and number of degrees of freedom equal to N. Figure 1.3 is a table with values of $t(N, \alpha)$ for various N and α.

The above expression claims that *the probability that the "true" mean value belongs to the interval* $[M_x(N) \pm \Delta^M(N, \alpha)]$ *is equal to* $1-2\alpha$.

The confidence interval for the standard deviation is:

$$P\{S_x(N) - \Delta^S(N, \alpha) \leq \sigma \leq S_x(N) + \Delta^S(N, \alpha)\} = 1 - 2\alpha$$

where $\Delta^S(N, \alpha) = t(N - 1, \alpha) \dfrac{S_x(N)}{\sqrt{2N}}$

The notations are,

$M_x(N)$ and $S_x(N)$ are mean value and standard deviation estimated using N observations,

$\sigma = S_x(\infty)$ is the "true" standard deviation,

$\Delta^S(N, \alpha)$ is width of the confidence interval that depends on the number of observations N and the significance level α.

$t(N-1, \alpha)$ is the t-distribution or Student distribution.

Tabulated Students t-Distribution Law The above expression establishes that *the probability that the "true" standard deviation value belongs to the interval* $[S_x(N) \pm \Delta^S(N, \alpha)]$ *is equal to* $1-2\alpha$.

The analysis of the formulae for confidence intervals and the knowledge of properties of t-distribution indicate that the confidence interval widens as standard

v	α									
	.40	.25	.10	.05	.025	.01	.005	.0025	.001	.0005
1	.325	1.000	3.078	6.314	12.706	31.821	63.657	127.32	318.31	636.62
2	.289	.816	1.886	2.920	4.303	6.965	9.925	14.089	23.326	31.598
3	.277	.765	1.638	2.353	3.182	4.541	5.841	7.453	10.213	12.924
4	.271	.741	1.533	2.132	2.776	3.747	4.604	5.598	7.173	8.610
5	.267	.727	1.476	2.015	2.571	3.365	4.032	4.773	5.893	6.869
6	.265	.718	1.440	1.943	2.447	3.143	3.707	4.317	5.208	5.959
7	.263	.711	1.415	1.895	2.365	2.998	3.499	4.029	4.785	5.408
8	.262	.706	1.397	1.860	2.306	2.896	3.355	2.833	4.504	5.041
9	.261	.703	1.383	1.833	2.262	2.821	3.250	3.690	4.297	4.781
10	.260	.700	1.372	1.812	2.228	2.764	3.169	3.581	4.144	4.587
11	.260	.697	1.363	1.796	2.201	2.718	3.106	3.497	4.025	4.437
12	.259	.695	1.356	1.782	2.179	2.681	3.055	3.428	3.930	4.318
13	.259	.694	1.350	1.771	2.160	2.650	3.012	3.372	3.852	4.221
14	.258	.692	1.345	1.761	2.145	2.624	2.977	3.326	3.787	4.140
15	.258	.691	1.341	1.753	2.131	2.602	2.947	3.286	3.733	4.073
16	.258	.690	1.337	1.746	2.120	2.583	2.921	3.252	3.686	4.015
17	.257	.689	1.333	1.740	2.110	2.567	2.898	3.222	3.646	3.965
18	.257	.688	1.330	1.734	2.101	2.552	2.878	3.197	3.610	3.922
19	.257	.688	1.328	1.729	2.093	2.539	2.861	3.174	3.579	3.883
20	.257	.687	1.325	1.725	2.086	2.528	2.845	3.153	3.552	3.850
21	.257	.686	1.323	1.721	2.080	2.518	2.831	3.135	3.527	3.819
22	.256	.686	1.321	1.717	2.074	2.508	2.819	3.119	3.505	3.792
23	.256	.685	1.319	1.714	2.069	2.500	2.807	3.104	3.485	2.767
24	.256	.685	1.318	1.711	2.064	2.492	2.797	3.091	3.467	3.745
25	.256	.684	1.316	1.708	2.060	2.485	2.787	8.078	3.450	3.725
26	.256	.684	1.315	1.706	2.056	2.479	2.779	3.067	3.435	3.707
27	.256	.684	1.314	1.703	2.052	2.473	2.771	3.057	3.421	3.690
28	.256	.683	1.313	1.701	2.048	2.467	2.763	3.047	3.408	2.674
29	.256	.683	1.311	1.699	2.045	2.462	2.756	3.308	3.396	3.659
30	.256	.683	1.310	1.697	2.042	2.457	2.750	3.030	3.385	3.646
40	.255	.681	1.303	1.648	2.021	2.423	2.704	2.971	3.307	3.551
60	.254	.679	1.296	1.671	2.000	2.390	2.660	2.915	3.232	3.460
120	.254	.677	1.289	1.658	1.980	2.358	2.617	2.860	3.160	3.373
∞	.253	.674	1.282	1.645	1.960	2.326	2.576	2.807	3.090	3.291

Source: Adapted with permission from *Biometrika Tables for Statisticians*, Vol. 1, 3rd ed., 1966, by E. S. Pearson and H. O. Hartley, Cambridge University Press, Cambridge.

Fig. 1.3 t-Distribution table. https://www.safaribooksonline.com/library/view/introduction-to-lin ear/9780470542811/22_app-a.html

deviation S_x increases and/or number of observations N decreases, or the required degree of certainty $P = 1-2\alpha$ increases. It can be seen that in order to increase the estimation accuracy without the sacrifice of certainty (or reliability of the estimation) one has only one approach: increasing the number of observations.

Example 1.5 Mean value and standard deviation of a random variable estimated using 43 measurements are: $M_x(43) = 13.6$ and $S_x(43) = 3.2$; define the confidence intervals for these values with the confidence of 95 % and the confidence of 99 %, i.e. define the intervals that would contain the "true" values of these parameters with probability of .95 and with probability of .99.

These confidence probabilities correspond to significance levels of $\alpha_1 = (1-.95)/2 = .025$ and $\alpha_2 = (1-.99)/2 = .005$. From the table in Fig. 1.3, $t(43,.025) \approx t(40,.025) \approx 2.021$, $t(43,.005) \approx t(40,.005) \approx 2.704$.

Therefore,

$$P\left\{13.6 - \frac{2.021 \times 3.2}{\sqrt{43}} \le \mu \le 13.6 + \frac{2.021 \times 3.2}{\sqrt{43}}\right\} = P\{12.61 \le \mu \le 14.59\} = .95 \text{ and}$$

$$P\left\{3.2 - \frac{2.021 \times 3.2}{\sqrt{2 \times 43}} \le \sigma \le 3.2 + \frac{2.021 \times 3.2}{\sqrt{2 \times 43}}\right\} = P\{2.5 \le \sigma \le 3.9\} = .95$$

Also,

$$P\left\{13.6 - \frac{2.704 \times 3.2}{\sqrt{43}} \le \mu \le 13.6 + \frac{2.704 \times 3.2}{\sqrt{43}}\right\} = P\{12.28 \le \mu \le 14.92\} = .99 \text{ and}$$

$$P\left\{3.2 - \frac{2.704 \times 3.2}{\sqrt{2 \times 43}} \le \sigma \le 3.2 + \frac{2.704 \times 3.2}{\sqrt{2 \times 43}}\right\} = P\{2.27 \le \sigma \le 4.13\} = .99$$

Example 1.6 How many measurements of the iron ore density (Lb/cm^3) should be performed to achieve the accuracy of the result of at least .01 (Lb/cm^3) with the confidence of 95 %? Note that this analysis is performed by averaging results of particular measurements and should be viewed as the estimation of the mean value of a random variable, consequently, the estimation error for 95 % confidence is,

$$\Delta^M(N, \alpha) = \Delta^M(N, .025) = t(N, .025)\frac{S_x(N)}{\sqrt{N}} = .01 \text{ (abs. units)}$$

It could be seen that finding the solution requires the knowledge of S_x that could be obtained by an auxiliary experiment. Assume that standard deviation of the analysis error was defined by conducting 25 experiments, $S_x(25) = 0.037$. Now the following computations must be performed:

$$t(25, .025)\frac{S_x(25)}{\sqrt{N}} = 2.06\frac{.037}{\sqrt{N}} = \frac{.07622}{\sqrt{N}} = .01 \Rightarrow \frac{.07622}{.01} = \sqrt{N} \Rightarrow N = 58$$

It could be seen that the expected solution implies 58 measurements, therefore t-distribution $t(58, .025) = 2.0$ should be utilized in the above calculation. Therefore,

$$t(60, .025)\frac{S_x(25)}{\sqrt{N}} = 2.0\frac{.037}{\sqrt{N}} = \frac{.074}{\sqrt{N}} = .01 \Rightarrow \frac{.074}{.01} = \sqrt{N} \Rightarrow N = 55$$

Exercise 1.1

Problem 1 According to the manufacturer, the probability of failure of a machine tool during its first 1000 h of operation is .083. The available statistics feature 56 cases of the machine tool failure during the first 1000 h of operation. In 16 out of these 56 cases, prior to the failure excessive vibration of the machine tool was observed, and in 7 cases overheating took place. Reassess the probability of failure of the machine tool during the first 1000 h of its operation knowing that excessive vibration has been observed once. Reassess this probability for the same machine tool based on the occurrence of temporary overheating. Note that under normal operational conditions the probability of overheating is .1 and the probability of excessive vibration is .05.

Problem 2 Consider Example 1.1 of the class notes. Define the probability of having poor quality of the product if at 10 am a short-time electrical failure was reported, at 2 pm poor quality of the raw material was observed, and later at 4 pm a minor equipment malfunction took place.

Problem 3 During the automatic assembly process, an electronic component must be placed on the board with the accuracy of $\pm.001$ inches, otherwise the board will be rejected by the quality controller. Under the existing technology positioning accuracy is characterized by the standard deviation of the positioning error, $\sigma = .0009$ inches. The improved controls may result in the reduction of the positioning error to $\sigma = .0003$ inches. Evaluate the productivity increase in % due to improved controls. (Assume normal distribution.)

Problem 4 According to a conducted study, an average student needs 2.5 min to read one page of the class notes with the standard deviation $\sigma = 0.6$ min. How much time is required for 90 % of students to read one page?

Problem 5 The findings given in the previous problem are based on the test conducted on 17 students. (a) Evaluate the confidence interval for the mean value and standard deviation featured in the problem. (b) How many students should be tested to double the accuracy of the estimation of the mean value? (c) How many students should be tested to double the accuracy of the estimation of the standard deviation? Perform these calculations twice: for 95 % confidence and 90 % confidence.

1.3   Systems of Random Variables

Consider a group of 3 random variables, $x(i)$, $y(i)$, $z(i)$, $i = 1,2,\ldots,N$, where $i = 1,2,\ldots,N$ is the realization index. How can the general properties of this group of 3 random variables be extracted?

1. Find their Min and Max values: $[X_{MIN}, X_{MAX}]$, $[Y_{MIN}, Y_{MAX}]$, $[Z_{MIN}, Z_{MAX}]$
2. Divide the above intervals into L subintervals, thus resulting in three steps, $\Delta_X = [X_{MIN} - X_{MAX}]/L$, $\Delta_Y = [Y_{MIN} - Y_{MAX}]/L$, and $\Delta_Z = [Z_{MIN} - Z_{MAX}]/L$
3. Compute numbers N_{KJM}, equal to the number of realizations $[x(i), y(i), z(i)]$ such that
   $X_{MIN} + (K-1)\Delta_X \leq x(i) < X_{MIN} + K\Delta_X,$
   $Y_{MIN} + (J-1)\Delta_Y \leq y(i) < Y_{MIN} + J\Delta_Y,$
   $Z_{MIN} + (M-1)\Delta_Z \leq z(i) < Z_{MIN} + M\Delta_Z$
   for every K, J, M $= 1, 2, 3, \ldots, L$

4. Compute frequencies $F_{KJM} = N_{KJM}/N$, K, J, $M = 1,2,3,\ldots,L$ of the multi-dimensional histogram
5. Assume $N \to \infty, N_{KJM} \to \infty, \Delta_X \to 0, \Delta_Y \to 0, \Delta_Z \to 0$, then the histogram turns into a 3-dimensional probability density function, $F(x,y,z,\mu_x,\mu_y,\mu_z,\sigma_x,\sigma_y,\sigma_z,r_{XY},$ $r_{XZ},r_{YZ})$, representing the distribution law, where

μ_x,μ_y, μ_z and $\sigma_x, \sigma_y, \sigma_z$ are mean values and standard deviations of the respective variables representing their individual distribution laws,

r_{XY}, r_{XZ}, r_{YZ} are parameters known as correlation coefficients representing interrelation between individual variables.

In the most practical applications we are dealing with the normal distribution law.

Correlation Coefficients r_{XY}, r_{XZ}, r_{YZ} could be estimated according to the formula that defines what is known as a normalized correlation coefficient

$$r_{XY} = r_{XY}(N) = \frac{1}{N \cdot S_X \cdot S_Y} \sum_{i=1}^{N} [x(i) - M_X] \cdot [y(i) - M_Y]$$

where M_X, M_Y, S_X, S_Y are estimates of mean values and standard deviations of particular variables. Note, that the *normalized correlation coefficient* does not exceed 1, by its absolute value, i.e. $-1 \leq r_{XY} \leq 1$. It represents the extent of the linear relationship between random variables x and y, not a functional relationship, but a tendency, i.e. the relationship that may or may not manifest itself at any particular test but could be observed on a large number of tests.

Note that

$$R_{XY} = R_{XY}(N) = \frac{1}{N} \sum_{i=1}^{N} [x(i) - M_X] \cdot [y(i) - M_Y]$$

– this is just a *correlation coefficient* (not normalized)

Confidence Interval for Correlation Coefficients The following expression defines the interval that with a particular probability that contains the "true" value of the correlation coefficient:

$$P\left[r_{XY} - \Delta^R(\alpha, N) \leq r_{XY}^{TRUE} \leq r_{XY} + \Delta^R(\alpha, N)\right] = 1 - 2\alpha$$

where $\Delta^R(\alpha, N) = t(\alpha, N - 1)\dfrac{1 - r_{XY}^2}{\sqrt{N}}$ and

$t(\alpha, N-1)$ is t-distribution value defined for significance level α and number of degrees of freedom $N-1$.

It is said that the estimate of correlation coefficient is statistically significant if $|r_{XY}| \geq \Delta^R(\alpha, N)$. Indeed, the correlation coefficient could be only positive or only negative therefore the confidence interval of a statistically-significant normalized correlation coefficient cannot include both positive and negative values.

Example 1.7 Estimated correlation coefficient between the MPG value of Nissan Pathfinder and outdoor temperature was obtained using 20 available records, $r = 0.12$. Does this indicate that the correlation between these two random variables exists?

Assume significance level $\alpha = .025$, then t$(.025, 19) = 2.093$, then

$$D = 2.093 \cdot [1 - .0144]/4.47 = 0.46$$

$0.46 > 0.12$, therefore, with 95 % confidence the r value is statistically insignificant.

Conditional Distributions Consider a group of 2 random variables, $x(i)$ and $y(i)$, $i = 1,2,\ldots,N$, where $i = 1,2,\ldots,N$ is the realization index. Let us investigate if there is a trend-type dependence of random variable $y(i)$ on random variable $x(i)$

1. Find Min and Max values of these variables: $[X_{MIN}, X_{MAX}]$, $[Y_{MIN}, Y_{MAX}]$
2. Divide the above intervals into L subintervals, thus resulting in steps,

$$\Delta_X = [X_{MIN} - X_{MAX}]/L \text{ and } \Delta_Y = [Y_{MIN} - Y_{MAX}]/L$$

3. From the original array $x(i)$, $y(i)$, $i = 1,2,\ldots,N$, select only the realizations $[x(i), y(i)]$ such that

$$X_{MIN} + (K - 1)\Delta_X \leq x(i) < X_{MIN} + K\Delta_X,$$

Assume that the total number of such realizations is N_K

4. Obtain histogram for random variable $y(i)$ from the above array of N_K observations by

4a. Computing number of realizations, $N^M{}_K$ such that

$$Y_{MIN} + (M - 1)\Delta_Y \leq y(i) < Y_{MIN} + M\Delta_Y$$

for every $M = 1,2,3,\ldots,L$

4b. Compute frequencies $F^M{}_K = N^M{}_K/N_K$, $M = 1,2,3,\ldots,L$ of the multi-dimensional histogram

(Note that the above histograms in Fig. 1.4 for variable $y(i)$ are built only for those values $y(i)$ when corresponding $x(i)$ values fall within interval $[X_{MIN} + (K-1)\Delta_X, <X_{MIN} + K\Delta_X])$

5. Assume $N \to \infty$, $N_K \to \infty$, $N^M{}_K \to \infty$, $\Delta_X \to 0$, $\Delta_Y \to 0$, then the histogram turns into a 1-dimensional probability density function, $P(y,\mu_Y,\sigma_Y)$, representing the distribution law of variable $y(i)$ obtained under the assumption that corresponding values of $x(i)$ satisfy some particular conditions, i.e. $P(y,\mu_Y,\sigma_Y) = P(y,\mu_Y,\sigma_Y/x)$ is a

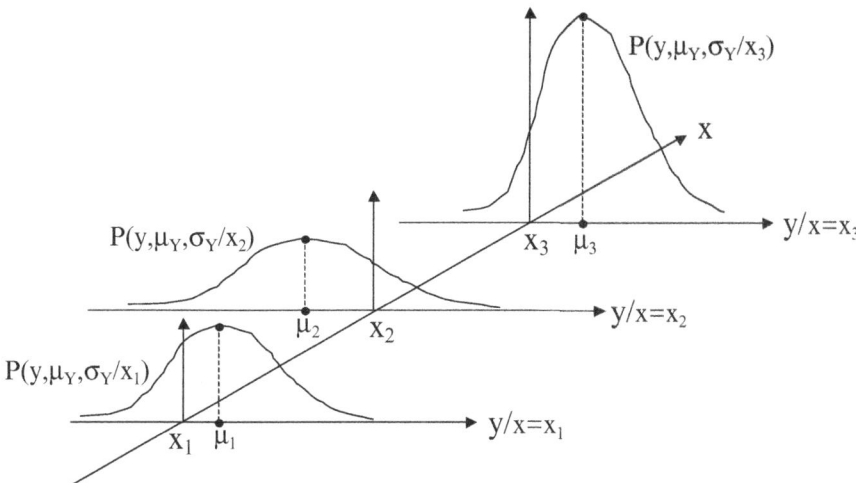

Fig. 1.4 Probability distributions of y for various x

probability density representing the distribution law of random variable y subject to variable x. This concept is illustrated by the above Fig. 1.4, note that both the mean value and variance (the spread of the distribution curve) of variable y change subject to numerical values of the related variable x.

One could expect that $P(y,\mu_Y,\sigma_Y) = P(y,\mu_Y,\sigma_Y/x)$ is a normal distribution, its dependence on x manifests itself as the dependence of its parameters, μ_Y and σ_Y, i.e. $\mu_Y = \mu_Y(x)$ and $\sigma_Y = \sigma_Y(x)$. These relationships are known as "conditional mean value" and "conditional standard deviation." In reality, standard deviation seldom depends on other variables, but the *conditional mean value* has a very important role.

Example 1.8 Given three arrays of observations of two random variables, $x(i)$ and $y(i)$, $i = 1,2,3,\ldots$ Based on the value of variable $x(i)$ the observations are divided into three groups, and mean values and standard deviations of the corresponding values of variable $y(i)$, i.e. M_Y and S_Y, are calculated within these groups, see the table below:

	x-range	Number of observations	Mean value M_Y	Standard deviation S_Y
Group 1	$5.0 \le x(i) < 7.0$	375	10.25	1.73
Group 2	$7.0 \le x(i) < 9.0$	633	10.67	1.82
Group 3	$9.0 \le x(i) \le 11.0$	592	10.91	1.91

Determine if this information indicates that variable $y(i)$ depends on variable $x(i)$.

Solution It could be concluded that variable $y(i)$ depends on variable $x(i)$ if the differences between mean values M_Y and/or standard deviation values S_Y in particular groups are statistically significant, i.e. exceed the half-width of the corresponding confidence intervals.

Assume the significance level $\alpha = .025$ then t-distribution is $t(\infty, .025) = 1.96$. The relevant quantities are as follows:

	$\Delta^M(N, \alpha) = t(N, \alpha)\dfrac{S_x(N)}{\sqrt{N}}$	$\Delta^S(N, \alpha) = t(N, \alpha)\dfrac{S_x(N)}{\sqrt{2N}}$
Group 1	0.175	.124
Group 2	0.142	.1
Group 3	0.154	.109

The analysis of the above result indicates that the differences between mean values,

Group 1/Group 2

$$\Delta_{12} = 10.76 - 10.25 = .42 > .175$$

and Group 2/Group 3

$$\Delta_{23} = 10.91 - 10.67 = .24 > .142$$

exceed the half-widths of the appropriate confidence intervals and therefore are statistically significant with the significance level of .025. At the same time, differences between standard deviations, Group 1/Group 2

$$\delta_{12} = 1.73 - 1.82 = .09 < .124$$

and Group 2/Group 3

$$\delta_{23} = 1.91 - 1.82 = .09 < .10$$

are less than the half-widths of the confidence intervals and therefore are statistically insignificant with the significance level of .025. This finding should be summarized as follows: with probability 95 % the mean value of variable y is affected by variable x, however there is no evidence that standard deviation of variable y depends on the value of variable x.

Regression Equation How to quantify the existing trend-type relationship between variables y and x? Unlike a functional relationship, for any value $x = x(i)$ a particular value of variable $y = y(i)$ is defined through a function $y = \varphi(x)$. A trend does not imply that for any $x(i)$ a specific $y(i)$ is prescribed. However, a trend manifests itself by a functional relationship between value $x(i)$ and the mean values μ_Y of the random variable $y(i)$corresponding to this $x(i)$, i.e. $\mu_Y = \varphi(x)$. This is known as a *regression equation*. There is another way to define a regression equation: $y = E\{y/x\}$ which is the conditional mean value of y subject to $x.E\{.../...\}$ is a symbol of conditional mathematical expectation. A regression equation does allow computation of the mean value of random variable y for any particular value $x(i)$, but what about the specific value $y(i)$ that will be observed in conjunction with $x(i)$? It should be clear that $y(i)$ will be different from $E\{y/x(i)\}$ and the expected difference depends on the variability of y represented by its variance, σ_Y^2, that may or may not depend on $x(i)$. Unlike correlation, regression is suitable for nonlinear relationships.

A trend-type relationship may exist between random variable y and a number of random variables, $x_1, x_2, x_3,\ldots, x_n$ in the sense that the mean value of y could be expressed as a function of these variables, $\mu_Y = \varphi(x_1, x_2, x_3,\ldots, x_n)$. In this case we are dealing with a *multiple regression equation.*

Regression equations are commonly used as mathematical models of manufacturing processes; therefore, development of such equations will be presented in the next chapter.

Correlation Analysis Correlation analysis is the analysis of stochastic (trend-type) linear relationships between random variables $x_1, x_2, x_3, \ldots, x_n$. It includes:

Computation of the correlation coefficient for every combination of two variables,

$$r_{ij} = \frac{1}{N \cdot S_i \cdot S_j} \sum_{k=1}^{N} [x_i(k) - M_i] \cdot [x_j(k) - M_j]$$

Computation of the correlation matrix,

$$\begin{bmatrix} r_{11} & r_{12} & \cdots & r_{1n} \\ r_{21} & r_{22} & \cdots & r_{2n} \\ \cdots & \cdots & \cdots & \cdots \\ r_{n1} & r_{n2} & \cdots & r_{nn} \end{bmatrix}$$

Computation of the multiple correlation coefficient,

$$R_{y,x_1,x_2,x_3} = \sqrt{\frac{\mathrm{Det}\begin{bmatrix} r_{11} & r_{12} & r_{13} & r_{1y} \\ r_{21} & r_{22} & r_{23} & r_{2y} \\ r_{31} & r_{32} & r_{33} & r_{3y} \\ r_{y1} & r_{y2} & r_{y3} & r_{yy} \end{bmatrix}}{\mathrm{Det}\begin{bmatrix} r_{11} & r_{12} & r_{13} \\ r_{21} & r_{22} & r_{23} \\ r_{31} & r_{32} & r_{33} \end{bmatrix}}} \qquad \text{Note that } 0 \le R_{y,x_1,x_2,x_3} \le 1$$

Where r_{jk} and r_{jy}—are normalized correlation coefficients between variables x_j and x_k and between x_j and y.

1.4 Random Processes

A random process could be viewed as a continuous function of time $y = y(t)$ that at any particular moment of time, t^*, has a random value, i.e. $y^* = y(t^*)$ is a random variable characterized by its specific distribution law. Recording a random process over some period of time would result in a graph as the one shown below in Fig. 1.5. It is said that the graph features a *realization of the random process $y(t)$*.

However the same random process repeatedly initiated under the same conditions would result in many different realizations that in combination constitute an *ensemble*, see below in Fig. 1.6.

The broken line in Fig. 1.6, representing time $t = t^*$, is known as the cross-section of the random process $y(t)$. It could be seen that in the cross-section multiple realizations form a combination of numerical values of random variables, i.e. when the time argument is fixed, $y(t^*)$ is a random variable with all previously described properties and characteristics.

Due to the proliferation of computers, we should expect to deal with discretized random processes represented by a sequence of random variables *attached to the*

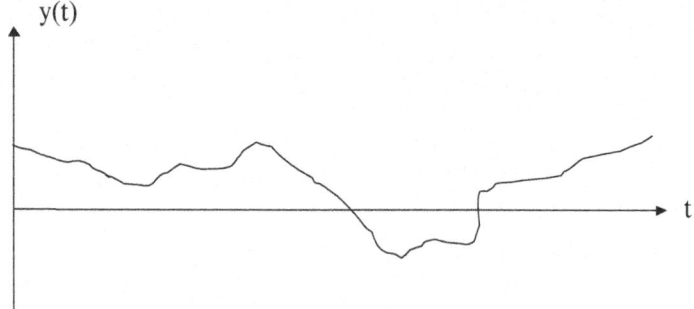

Fig. 1.5 Random process performance over time

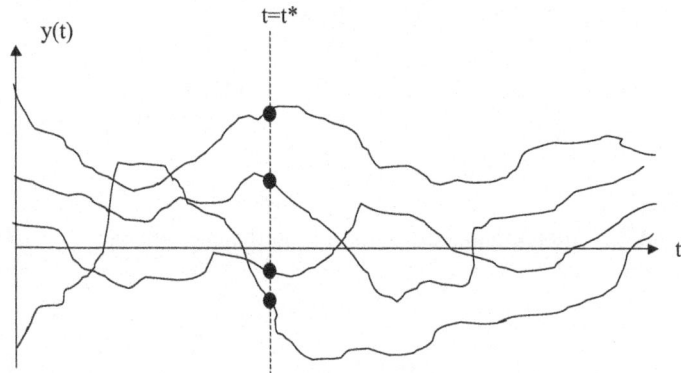

Fig. 1.6 Multiple iterations of random process

time axis, $y(\Delta t)$, $y(2 \cdot \Delta t)$, ..., $y(i \cdot \Delta t)$, ..., $y(N \cdot \Delta t)$, or just $y(1)$, $y(2)$, ..., $y(i)$, ..., $y(N)$, where $i = 1,2,...,N$ is the discrete-time index, Δt is the time step, and $N\Delta t$ is the entire period of observation. It should be emphasized that "attached to the time axis" is the key to distinguishing between a random variable and a random process. While sequencing of the observations of a random variable is not important, the "natural" sequencing of numbers representing a random process is crucial for the analysis of the phenomenon represented by this random process.

A discretized ensemble of the realizations of a random process could be represented by a square table, where rows represent particular realizations and columns represent particular cross-sections (discrete-time values):

Realization index	Discrete-time values						
	$t = \Delta t$	$t = 2 \cdot \Delta t$	$t = 3 \cdot \Delta t$	...	$t = i \cdot \Delta t$	...	$t = N \cdot \Delta t$
1	$y(1,1)$	$y(1,2)$	$y(1,3)$	...	$y(1,i)$	...	$y(1,N)$
2	$y(2,1)$	$y(2,2)$	$y(2,3)$	...	$y(2,i)$	...	$y(2,N)$
...	...	...	...	...	...	...	...
j	$y(j,1)$	$y(j,2)$	$y(j,3)$	...	$y(j,i)$	...	$y(j,N)$
...	...	...	...	...	...	...	...
M	$y(M,1)$	$y(M,2)$	$y(M,3)$	...	$y(M,i)$	...	$y(M,N)$

The approach to analyzing statistical properties of a random process is similar to the one suggested for a random variable: first a histogram is built provided that $N \gg 1$ and $M \gg 1$ and eventually a distribution law is established. However, immediate questions arise:

1. Should this distribution law be established for a realization of the process (one of the rows of the table) or for a cross-section (one of the columns of the table)?
2. Is it necessary to establish an individual distribution law for every cross-section (column) of the process?

Answers to these questions reflect fundamental properties of the random process:

1. A random process is called *ergodic* if a distribution law established for one of its realizations is identical to the one established for its cross-section. Otherwise the process is said to be *non-ergodic*.
2. A random process is called *stationary* if a distribution law established for a cross-section is independent of the cross-section. This implies that statistical characteristics of the process are time-invariant. Otherwise it is said that the process is *non-stationary* or has a *parameter drift*.
3. Any ergodic process is stationary, but not every stationary process is ergodic.

Most realistic random processes are ergodic and the normal distribution law is suitable for their description. It is also known that most non-stationary random processes are non-stationary only in terms of the mean value, i.e. $\mu = \mu(t)$, but their standard deviation σ is constant.

A random process may involve one or more variables. A multi-variable distribution law characterizes a multi-variable random process.

Autocorrelation Function Consider two cross-sections of a random process separated by time interval τ, i.e. $y(t_1)$ and $y(t_1 + \tau)$. It is known that $y(t_1)$ and $y(t_1 + \tau)$ are two random variables that may or may not be correlated. It is clear that when $\tau = 0$, $y(t_1 + \tau)$ simply repeats $y(t_1)$ and $y(t_1 + \tau) = y(t_1)$ is just a linear functional relationship between these two random variables. Intuitively, due to variability of the random process the resemblance between $y(t_1 + \tau)$ and $y(t_1)$ decreases with the increase of time interval τ. Consequently, correlation between these two random variables is expected to exist, to be positive, to decrease with the increase of τ, and to approach zero as $\tau \to \infty$. The rate of decrease of this correlation represents important properties, primarily inertia, of the underlying physical phenomena. An autocorrelation function is a numerical tool for the analysis of the correlation between any two cross sections of a random process, $y(t)$ and $y(t + \tau)$. It is defined as a function of time interval τ, represented by the integer number n, and is estimated as follows:

$$r_Y(\tau) = r_Y(n) = \frac{1}{(N - n) \cdot S_Y^2} \sum_{i=1}^{N-n} [y(i) - M_Y] \cdot [y(i + n) - M_Y],$$

$$n = 0, 1, 2, \ldots, n^*; \quad n^* < N$$

where
$y(i)$ is a discrete-time value of the available realization of the random process,
M_Y and S_Y^2 are estimated mean value and variance of the random process $y(t)$,
$n = \tau/\Delta t$ is the time shift representing interval τ by an integer number of time steps Δt,
$N \gg 1$ is the total number of data points of the realization of the random process.

One should realize the particular values $r_Y(n)$, $n = 0,1,2,\ldots$ are nothing but normalized correlation coefficients and, as such, could be statistically significant or insignificant. Recall that the significance condition of a correlation coefficient $r_Y(n)$ estimated using N-n data points is defined as $|r_Y(n)| > \Delta^R$ where $\Delta^R = t$ $(\alpha, N - n) \frac{1 - r_Y(n)^2}{\sqrt{N-n}}$ is the width of the confidence interval of the estimate and $t(N$-n, $\alpha)$ is the t-distribution value defined for significance level α and the number of degrees of freedom N-n. It could be seen that $\Delta^R = \Delta^R(n)$ and analysis of the above formula indicates that this is an increasing function of n.

Figure 1.7 depicts an estimated autocorrelation function of a random process $y(t)$ and the width of the confidence intervals of its values obtained for $N \gg 1$.

It could be seen that the estimated correlation function is statistically significant only for $n \leq n^{\text{CORR}}$. Time interval $\tau^{\text{CORR}} = \Delta t \cdot n^{\text{CORR}}$ is known as the correlation time of the random process $y(t)$ and sometimes is called the *memory* of this process.

Fig. 1.7 Random function autocorrelation

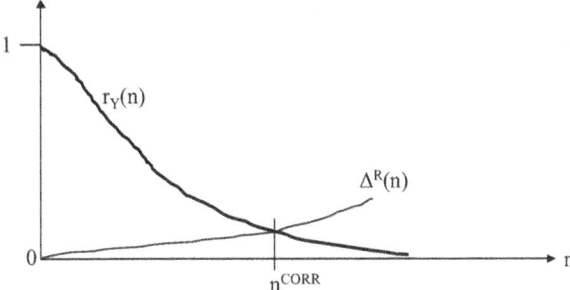

Note that the number of values of the correlation function that are being calculated is $n^* \approx N/4$. However, it is meaningless to go beyond the limit of significance n^{CORR}.

Example 1.9 A computer-based monitoring system for an industrial power generator is being designed. The system will monitor generator variables, submit them for analysis and record data in a permanent database. Determine the required time interval for analyzing and recording data if the following autocorrelation function of one of the variables is available.

n	0	1	2	3	4	5	6	7	8	9	10
$r_X(n)$	1.0	.95	.81	.61	.23	.09	.03	.01	.006	.008	.007

Note that this correlation function was obtained using 1000 data points taken with the time step $\Delta t = 60$ s.

Solution First, let us determine the correlation time n^{CORR} by establishing the half-width of the confidence interval for every available value of the correlation function using the formula $\Delta^R = t(\alpha, N - n) \frac{1 - r_Y(n)^2}{\sqrt{N-n}}$ and assuming $\alpha = .025$. Note that according to the t-distribution table, the t-distribution value stays the same, $t(\infty, 0.025) = 1.98$, for all points. The results are shown in the table below.

n	0	1	2	3	4	5	6	7	8	9	10
$r_X(n)$	1.0	.95	.81	.61	.23	.09	.03	.01	.006	.008	.007
Δ^R	0.	.001	.003	.006	.010	.0103	.0105	.011	.0108	.0109	.011

It could be easily seen that at $n = 7$, $r_X(n) \approx \Delta^R$ and at $n > 7$ $r_X(n) < \Delta^R$ thus $n^{CORR} = 7$ and the correlation time for this process is 7 min. Consequently, the 7 min period should be recommended as a rational time period for analyzing and recording generator data. Indeed, having this time period under 7 min would result in analyzing and recording data that contains duplicating information. Having this time period greater than 7 min implies that some valuable information could be lost.

Cross-Correlation Function Cross-correlation function works exactly as the auto-correlation function, but it describes the correlation between two random processes (or two components of a multivariable random process), say $x(t)$ and $y(t)$. In order to simplify the entire discussion of cross-correlation functions let us assume that process $x(t)$ could be viewed as the input (stimuli) of some physical process, and $y(t)$ as its response.

Cross-correlation function is estimated as follows:

$$r_{XY}(n) = \frac{1}{(N-n) \cdot S_Y S_X} \sum_{k=1}^{N-n} [x(i) - M_X] \cdot [y(i+n) - M_Y], n = 0, 1, 2, \ldots, n^*; n < N$$

where

$x(i)$ and $y(i)$ are discrete-time values of the available realizations of the random
   processes,
M_X, M_Y and S_X, S_Y are estimated mean values and standard deviations of the random
   processes $x(t)$ and $y(t)$,
τ is the time distance between the cross-section of the process $x(t)$ and the cross-
   section of the process $y(t)$,
$n = \tau/\Delta t$ is the time interval τ represented by an integer number of time steps Δt,
   and
$N \gg 1$ is the total number of data points of the realization of the random processes
   x(t) and y(t).

Since particular values $r_{XY}(n)$, $n = 0,1,2,\ldots$ are nothing but normalized correlation coefficients they could be statistically significant or insignificant. The significance condition of an estimate $r_{XY}(n)$ is defined as $|r_{XY}(n)| > \Delta^R$ where $\Delta^R = t(\alpha, N - n) \frac{1 - r_{XY}(n)^2}{\sqrt{N-n}}$ is the width of the confidence interval of the estimate and $t(N-n, \alpha)$ is t-distribution value defined for significance level α and the number of degrees of freedom N-n.

Note that unlike auto-correlation function, cross-correlation function could be positive and negative but its absolute value cannot exceed 1.0. Typical configurations of cross-correlation functions and widths of the confidence intervals of their values, estimated for $N \gg 1$, are shown below in Fig. 1.8.

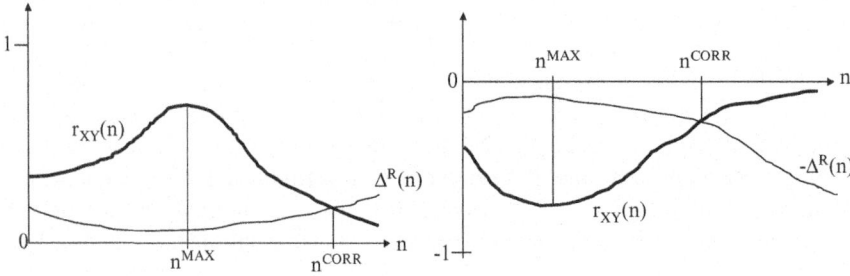

Fig. 1.8 Cross-correlation functions and confidence intervals

In addition to the limits beyond which estimated cross-correlation functions become statistically insignificant, n^{CORR}, a graph of a cross-correlation function has the maximum correlation point, n^{MAX}. Maximum correlation point indicates that the stimuli, $x(t)$, has maximum effect on the response, $y(t)$, not immediately but with some delay of $\tau^{MAX} = \Delta t \cdot n^{MAX}$, therefore, τ^{MAX} facilitates the assessment of the delay (transport lag) and/or inertia of the physical phenomenon represented by random processes $x(t)$ and $y(t)$. Indeed, auto- and cross-correlation functions present a technique to investigate and represent dynamics of an inertial system, however, they are not as versatile as transfer functions.

Example 1.10 A special study is conducted to determine the effect of carbon dioxide on a process taking place in a chemical reactor. During this study, the flow of carbon dioxide injected in the reactor (foot^3/s) and the percent of sulfur in the continuous flow of the end product were monitored with the sampling period of 20 s. A cross-correlation function between two resultant variables, the flow of carbon dioxide (treated as the stimuli) and the percent of sulfur (as the response) was obtained using 100 measurements, see table below.

n	0	1	2	3	4	5	6	7	8	9	10
$r_{XY}(n)$	.08	−.015	−.007	−.13	−.23	−.52	−.28	−.019	−.011	.06	.07

Give your conclusions on the carbon dioxide/percent of sulfur interaction on the basis of the available cross-correlation function.

Solution Let us investigate the statistical significance of the particular values of the cross-correlation function by computing their respective half-widths of the confidence intervals $\Delta^R = t(\alpha, N - n) \frac{1 - r_{XY}(n)^2}{\sqrt{N-n}}$ and assuming $\alpha = .025$. As in the previous example, $t(120, .025)$ could be utilized. The results are as follows,

n	0	1	2	3	4	5	6	7	8	9	10
$r_{XY}(n)$	.08	−.015	−.007	−.13	−.23	−.52	−.28	−.019	−.011	.06	.07
Δ^R	.197	.199	.200	.197	.191	.148	.188	.205	.206	.207	.208

Although cross-correlation function shows some positive and negative values, note that statistically significant values are only at $n = 4$, $n = 5$, and $n = 6$. Since these significant values are negative, one can conclude that carbon dioxide causes a reduction of the percent of sulfur in the end product. The second conclusion is that it takes approximately 100 s for carbon dioxide to have its effect on the percent of sulfur in the end product. Therefore, if further statistical studies of this effect would be conducted, the percent of sulfur data records must be appropriately aligned (shifted) with respect to the flow of carbon dioxide data.

Spectral Density Spectral density is another approach to address the variability of a random process. It implies that a random process could be represented by a combination of harmonics, i.e. particular sinusoidal signals defined by their frequencies, magnitudes and phases. Although theoretically the number of such harmonics is infinite, a random process is often dominated by relatively few harmonics. Spectral analysis is routinely performed by engineering software tools.

Exercise 1.2 Generate the following data arrays:

$$x(i) = 3 + 7 \cdot \sin(0.1 \cdot i) + 2 \cdot \sin(0.7 \cdot i) + 0.5 \cdot \sin(1.77 \cdot i) + 0.1 \cdot \sin(713 \cdot i), \ i = 1, \ldots, 300$$

$$y(i) = 13 + 17 \cdot \sin(i) + 2 \cdot \sin(.0137 \cdot i) + 0.8 \cdot \sin(6.77 \cdot i) + 0.4 \cdot \sin(7103 \cdot i) + 0.05 \cdot x(i),$$
$$i = 1, \ldots, 300$$

$$z(i) = -7 + \sin(0.5 \cdot i) + 2 \cdot \sin(3.7 \cdot i) + 0.05 \cdot \sin(1677 \cdot i) + 0.02 \cdot x(i) + 0.1 \cdot y(i),$$
$$i = 1, \ldots, 300$$

$$v(i) = x(i) + 5 \cdot \sin(0.02 \cdot i), \ i = 1, \ldots, 300$$

$$w(i) = z(i) + 11 \cdot \sin(0.02 \cdot i + 2.05), \ i = 1, \ldots, 300$$

Problem 1 Obtain correlation matrix for variables x, y, and z and evaluate statistical significance of every correlation coefficient for the significance level $\alpha = .005$.

Problem 2 Obtain multiple correlation coefficient R_{yxzv}

Problem 3 Investigate the possible effect of numerical values of variable x on the mean value and standard deviation of variable z by dividing the range of variable x into two equal subdivisions and analyzing corresponding values of characteristics of variable z.

Problem 4 Assume that variables $v(i)$ and $w(i)$ are discretized realizations of a random process. Obtain their cross-correlation function (treat $v(i)$ as the stimuli).

Problem 5 Use a standard software frequency analysis tool to investigate random process $x(i)$. Since the actual frequency composition of this signal is known to you, comment on the ability of this software tool to recover all harmonics of the signal.

Solutions

Exercise 1.1: Problem 1

The following probabilities can be extracted from the given information. Note that *fail* represents the event of a machine tool failure, *good* represents the event that there is no machine tool failure, *vibration* represents that there was excessive vibration, and *overheat* represents the event of overheating.

$$P(fail) = 0.083$$

$$P(good) = 0.917$$

$$P(vibration|fail) = \frac{16}{56} = 0.286$$

$$P(vibration|good) = 0.05$$

$$P(overheat|fail) = \frac{7}{56} = 0.125$$

$$P(overheat|good) = 0.1$$

Given these initial probabilities, the conditional probability that there was a failure given an observed vibration can be calculated as follows.

$$P(fail|vibration) = \frac{P(vibration|fail) \cdot P(fail)}{P(vibration|fail) \cdot P(fail) + P(vibration|good) \cdot P(good)}$$

$$P(fail|vibration) = \frac{0.286 \cdot 0.083}{0.286 \cdot 0.083 + 0.05 \cdot 0.917} = 0.34$$

Now, this conditional probability of failure is going to be used as the probability for failure in future calculations. Since the sum of all probabilities in a set must be one, the probability that the product is good must be 0.66. Now that the first event occurred and we have these new probabilities of failure, the probability of failure can be calculated given the next event.

$$P(fail|overheat) = \frac{P(overheat|fail) \cdot P(fail)}{P(overheat|fail) \cdot P(fail) + P(overheat|good) \cdot P(good)}$$

$$P(fail|overheat) = \frac{0.125 \cdot 0.34}{0.125 \cdot 0.34 + 0.1 \cdot 0.66} = 0.392$$

So, after both events occurred, the probability of a failure is 0.392.

Exercise 1.1: Problem 2

The following frequencies (probabilities) can be extracted from the Example data.

$$P(A|poor) = \frac{136}{879} = 0.155 \quad P(B|poor) = \frac{177}{879} = 0.201 \quad P(C|poor) = \frac{83}{879} = 0.094$$

$$P(A|good) = \frac{36}{4621} = 0.0078 \quad P(B|good) = \frac{81}{4621} = 0.0175 \quad P(C|good) = \frac{63}{4621} = 0.0136$$

$$P(poor) = 0.16 \quad P(good) = 0.84$$

Now, the probability of poor quality given event A at 10 am:

$$P(poor|A) = \frac{P(A|poor) \cdot P(poor)}{P(A|poor) \cdot P(poor) + P(A|good) \cdot P(good)}$$

$$P(poor|A) = \frac{0.155 \cdot 0.16}{0.155 \cdot 0.16 + 0.0078 \cdot 0.84} = 0.791$$

The new probabilities are:

$$P(poor) = 0.791 \; P(good) = 0.209$$

Now, the probability of poor quality given the consecutive event B at 2 pm:

$$P(poor|B) = \frac{P(B|poor) \cdot P(poor)}{P(B|poor) \cdot P(poor) + P(B|good) \cdot P(good)}$$

$$P(poor|B) = \frac{0.201 \cdot 0.791}{0.201 \cdot 0.791 + 0.0175 \cdot 0.209} = 0.978$$

The new probabilities are:

$$P(poor) = 0.978 \; P(good) = 0.022$$

Now, the probability of poor quality given the consecutive event C at 4 pm:

$$P(poor|C) = \frac{P(C|poor) \cdot P(poor)}{P(C|poor) \cdot P(poor) + P(C|good) \cdot P(good)}$$

$$P(poor|C) = \frac{0.094 \cdot 0.978}{0.094 \cdot 0.978 + 0.0136 \cdot 0.022} = 0.9986$$

Given the three sequential events, the probability that there was a poor quality of product became **99.86 %**.

Exercise 1.1: Problem 3

The probability of passing the requirements can be calculated for the original system as the area between two z-scores on a standard normal curve. The upper and lower bound z-scores are calculated as:

$$z_{pass}{}^+ = \frac{0.001 - 0}{0.0009} = 1.111 \quad z_{pass}{}^- = \frac{-0.001 - 0}{0.0009} = -1.111$$

The probability inside these bounds is 0.7335, or 73.4 %.

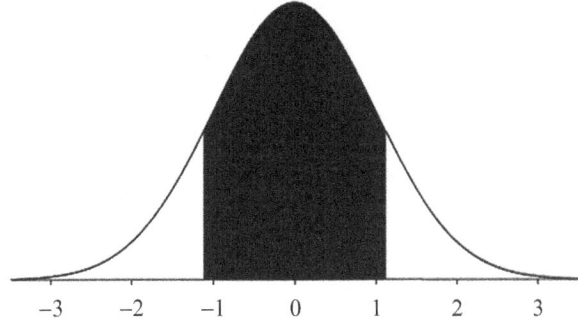

Now, to check the improvement from the controls added, the same process will be done with the statistical data for the controlled process.

$$z_{pass}{}^{+} = \frac{0.001 - 0}{0.0003} = 3.333 \qquad z_{pass}{}^{-} = \frac{-0.001 - 0}{0.0003} = -3.333$$

The probability inside these bounds is 0.9991, or 99.91 %.

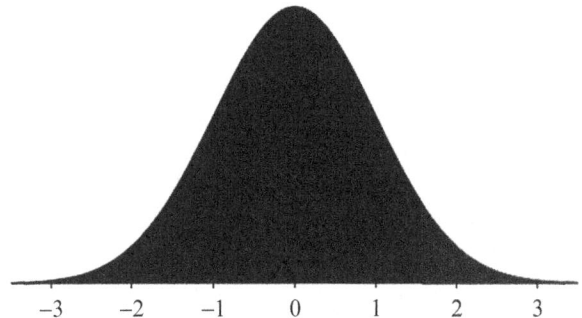

Given this information, we can see that the probability of passing the requirements jumped from 73.4 % to 99.91 % with the addition of controls. This means that there was a **25.61 %** increase in the success rate of this procedure from the introduction of controls.

Exercise 1.1: Problem 4

In this problem, we are looking for the amount of time sufficient for 90 % of students to complete the reading task. For this, we will look at a normal distribution with a mean of 2.5 and a standard deviation of 0.6 min. The z-value corresponding to 90 % probability under is 1.282.

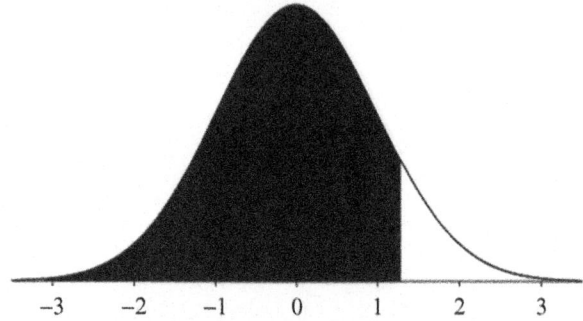

The reading time value associated with this z is 3.3 min. We can conclude that within 3.3 min, 90 % of students will be done reading one page.

Exercise 1.1: Problem 5

Part A

For a 90 % confidence interval and 17 students, the t-value for this calculation will be

$$mean : t(\alpha, N) = t(.05, 17) = 1.740$$
$$stddev : t(\alpha, N - 1) = t(.05, 16) = 1.746$$

The 90 % confidence interval for mean:

$$\Delta = t(\alpha, N) \cdot \frac{\sigma_N}{\sqrt{N}} = 1.740 \cdot \frac{0.6}{\sqrt{17}} = 0.253$$
$$P(\mu - \Delta \leq \mu_{TRUE} \leq \mu + \Delta) = 90\%$$
$$P(2.5 - .253 \leq \mu_{TRUE} \leq 2.5 + .253) = 90\%$$
$$P(2.247 \leq \mu_{TRUE} \leq 2.753) = 90\%$$

The 90 % confidence interval for standard deviation:

$$\Delta = t(\alpha, N - 1) \cdot \frac{\sigma_N}{\sqrt{2N}} = 1.746 \cdot \frac{0.6}{\sqrt{2 \cdot 17}} = 0.18$$
$$P(\sigma - \Delta \leq \sigma_{TRUE} \leq \sigma + \Delta) = 90\%$$
$$P(0.6 - .18 \leq \sigma_{TRUE} \leq 0.6 + .18) = 90\%$$
$$P(0.42 \leq \sigma_{TRUE} \leq 0.78) = 90\%$$

For a 95 % confidence interval and 17 students, the t-value for this calculation will be

$$mean : t(\alpha, N) = t(.025, 17) = 2.11$$
$$stddev : t(\alpha, N - 1) = t(.025, 16) = 2.12$$

The 95 % confidence interval for mean:

$$\Delta = t(\alpha, N) \cdot \frac{\sigma_N}{\sqrt{N}} = 2.11 \cdot \frac{0.6}{\sqrt{17}} = 0.307$$
$$P(\mu - \Delta \leq \mu_{TRUE} \leq \mu + \Delta) = 95\%$$
$$P(2.5 - .307 \leq \mu_{TRUE} \leq 2.5 + .307) = 95\%$$
$$P(2.193 \leq \mu_{TRUE} \leq 2.807) = 95\%$$

The 95 % confidence interval for standard deviation:

$$\Delta = t(\alpha, N - 1) \cdot \frac{\sigma_N}{\sqrt{2N}} = 2.12 \cdot \frac{0.6}{\sqrt{2 \cdot 17}} = .218$$
$$P(\sigma - \Delta \leq \sigma_{TRUE} \leq \sigma + \Delta) = 95\%$$
$$P(0.6 - .218 \leq \sigma_{TRUE} \leq 0.6 + .218) = 95\%$$
$$P(0.382 \leq \sigma_{TRUE} \leq 0.818) = 95\%$$

Part B

Doubling the accuracy with a 90 % confidence interval would require the following N.

$$\Delta = 0.253 \rightarrow \Delta_{NEW} = 0.1265$$

If we make our $N = 63$, we can observe a doubling in our accuracy by a halving of our interval width.

$$\Delta_{NEW} = 1.669 \cdot \frac{0.6}{\sqrt{63}} = 0.126$$

Doubling the accuracy of a 95 % confidence interval would require the following N.

$$\Delta = 0.307 \rightarrow \Delta_{NEW} = 0.1535$$

If we make our $N = 61$, we can observe a doubling in our accuracy by a halving of our interval width.

$$\Delta_{NEW} = 1.9996 \cdot \frac{0.6}{\sqrt{61}} = 0.1536$$

Part C

Doubling the accuracy for a 90 % confidence interval would require the following N.

$$\Delta = 0.18 \rightarrow \Delta_{NEW} = 0.09$$

If we make our $N = 60$, we can observe a doubling in our accuracy by a halving of our interval width.

$$\Delta_{NEW} = 1.746 \cdot \frac{0.6}{\sqrt{120}} = 0.095$$

Doubling the accuracy of a 95 % confidence interval would require the following N.

$$\Delta = 0.218 \rightarrow \Delta_{NEW} = 0.109$$

If we make our $N = 61$, we can observe a doubling in our accuracy by a halving of our interval width.

$$\Delta_{NEW} = 2.12 \cdot \frac{0.6}{\sqrt{122}} = 0.11$$

Exercise 1.2: Problem 1

The correlation matrix was calculated with the following configuration:

$$R_{xyz} = \begin{bmatrix} r_{xx} & r_{xy} & r_{xz} \\ r_{xy} & r_{yy} & r_{yz} \\ r_{xz} & r_{yz} & r_{zz} \end{bmatrix}$$

In which the correlation coefficient for two variables, x and y, is defined as

$$r_{xy} = \frac{1}{N \cdot \sigma_x \cdot \sigma_y} \cdot \sum_{n=1}^{N} [x(n) - \bar{x}] \cdot [y(n) - \bar{y}]$$

The correlation matrix for x, y, and z is:

$$R_{xyz} = \begin{bmatrix} 0.9967 & 0.0251 & 0.0719 \\ 0.0251 & 0.9967 & 0.6053 \\ 0.0719 & 0.6053 & 0.9967 \end{bmatrix}$$

Then, the statistical significance was evaluated by comparing the half-width if the confidence interval for the correlation coefficients to the coefficients themselves. The correlation coefficients were deemed significant of they were outside the confidence interval, meaning that the correlation coefficient was greater than the half-width of the interval. The half-widths of the intervals were calculated as

$$\Delta_{xy} = t(\alpha = .005, N = 300) \cdot \frac{1 - r_{xy}^2}{\sqrt{N}}$$

$\Delta_{xy} = 0.14956$ and r_{xy} is 0.02506, so the x-y correlation is not significant.
$\Delta_{xz} = 0.14888$ and r_{xz} is 0.071861, so the x-z correlation is not significant.
$\Delta_{yz} = 0.094818$ and r_{yz} is 0.60531, so the y-z correlation is significant.

Exercise 1.2: Problem 2

The multiple correlation coefficient $R_{y,xvz}$ is:

$$R_{y,xvz} = \sqrt{\frac{Det\begin{pmatrix} 0.9967 & 0.8146 & 0.0719 & 0.0251 \\ 0.8146 & 0.9967 & 0.0822 & 0.0577 \\ 0.0719 & 0.0822 & 0.9967 & 0.6053 \\ 0.0251 & 0.0577 & 0.6053 & 0.9967 \end{pmatrix}}{Det\begin{pmatrix} 0.9967 & 0.8146 & 0.0719 \\ 0.8146 & 0.9967 & 0.0822 \\ 0.0719 & 0.0822 & 0.9967 \end{pmatrix}}} = 0.7919$$

Exercise 1.2: Problem 3

First, the array of X was ordered from minimum to maximum value and then split into two equal parts. The associated Z values were split respectively into two equal-length sets. The mean value for Z for each set was calculated separately. The difference of these two mean values was compared to the half-widths of their confidence intervals.

The half-widths were calculated as:

$$\Delta_{Z1} = t(\alpha = .025, N = 50) \cdot \frac{\sigma_{Z1}}{\sqrt{N}} = 2.01 \cdot \frac{1.965}{\sqrt{50}} = 0.5586$$

$$\Delta_{Z2} = t(\alpha = .025, N = 50) \cdot \frac{\sigma_{Z2}}{\sqrt{N}} = 2.01 \cdot \frac{2.0875}{\sqrt{50}} = 0.5934$$

The difference in mean values of Z_1 and Z_2 is 0.3387, and the half-width of the 95 % confidence intervals for Z sets are 0.5586 and 0.5934. This indicates that there is no evidence that value of variable X has an effect on mean value of variable Z.

Exercise 1.2: Problem 4

The cross-correlation function is a function calculated with respect to discrete interval m that varies as m=0,1,2,...,$N/4$. The value of this function is:

$$r_{cross\_vw} = \frac{1}{(N - m) \cdot \sigma_v \cdot \sigma_w} \cdot \sum_{m=0}^{N/4} \sum_{i=1}^{N-m} [v(i) - \bar{v}] \cdot [w(i + m) - \bar{w}]$$

The resulting cross-correlation function is plotted below against m.

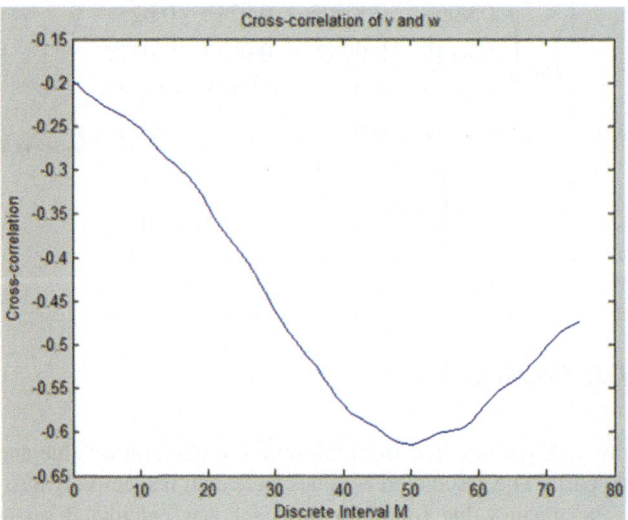

Exercise 1.2: Problem 5

A frequency analysis tool in MATLAB was used to break down the frequency spectrum of $x(i)$.

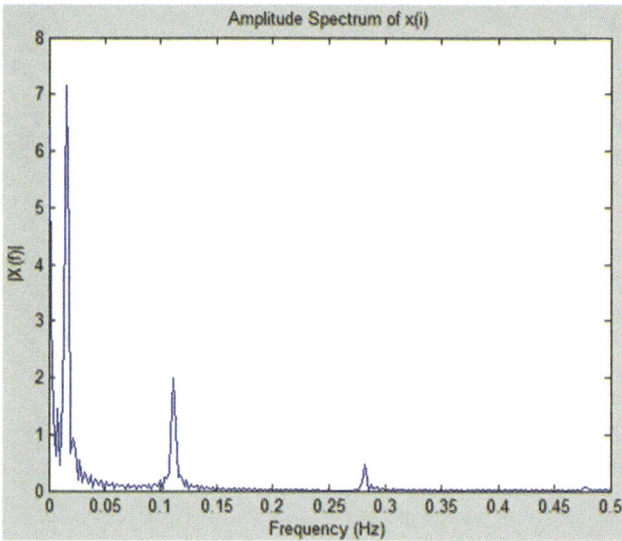

The detected peaks are consistent with the frequencies and magnitudes of the sinusoidal components of the signal:

- Peak at 0.016 Hz (0.10 radians/s) with amplitude 7,
- Peak at 0.11 Hz (0.70 radians/s) with amplitude 2
- Peak at 0.28 Hz (1.77 radians/s) with amplitude .5

Bibliography

1. Sam Kash Kachigan, Multivariate Statistical Analysis: A Conceptual Introduction, 2nd Edition, ISBN-13: 978-0942154917
2. Alvin C. Rencher, William F. Christensen, Methods of Multivariate Analysis 3rd Edition, Wiley, ISBN-13: 978-0470178966

Chapter 2
Mathematical Description of Manufacturing Processes

Mathematical description of a manufacturing process resulting in a mathematical model presents a basis for objective analysis, control, and optimization. Model-based process analysis includes prediction of the process outcome and its particular characteristics, "what if" analysis, and the evaluation of the effects of particular factors on the process. Mathematical models in the form of transfer functions are invaluable for the analysis and synthesis of control systems maintaining the desired operation of the process in spite of various disturbance factors. A mathematical model allows for the formalization of the process optimization problem, and serves as a "guinea pig" during the optimization search. Implemented in a simulation environment, a mathematical model presents an ideal testbed for the validation of the most advanced control and optimization schemes. In order to be usable, a mathematical model must be updated and valid. This section presents practical techniques for the development, validation, and updating of mathematical models utilizing statistical data.

2.1 Regression Analysis and the Least Squares Method

Regression analysis requires a data array and a configuration of the regression equation.

A data array includes synchronously recorded values of the input and the output variables of the process in question:

$$x_1(k), \; x_2(k), \; \ldots, \; x_m(k), \; \text{and } y(k),$$

where $k = 1, 2, \ldots, N$ is the discrete-time index or the realization index.

© Springer International Publishing Switzerland 2016   37
V.A. Skormin, *Introduction to Process Control*, Springer Texts
in Business and Economics, DOI 10.1007/978-3-319-42258-9_2

Model Configuration   A regression model is visualized as

$$y^{MOD}(k) = E\{y/x_1(k), x_2(k), \ldots, x_m(k)\},$$

where $E\{./.\}$ is the symbol of conditional mean value. It is good to realize that a regression model it is not a relationship between the output and the input variables but just a relationship between the mean value of the output and the inputs!

Case 1, linear relationship: $y^{MOD}(k) = \sum_{i=1}^{m} a_i x_i(k)$

Case 2, nonlinear relationship: $y^{MOD}(k) = \sum_{i=1}^{m} a_i x_i(k) + \sum_{i,\,j\,=\,1}^{m} a_{ij} x_i(k) x_j(k),$

Case 3, nonlinear relationship:

$$y^{MOD}(k) = \sum_{j=1}^{m} a_j \Phi_j[x_i(k), \ i = 1, 2, \ldots, m],$$

where $\Phi_j[x_i(k), \ i = 1, 2, \ldots, m]$ are nonlinear functions of $x_i(k), \ i = 1, 2, \ldots, m$

Least Squares Method   First, we will concentrate on linear models and take advantage of matrix–vector notations:

$$X_N = \begin{bmatrix} x_1(1) & x_2(1) & \ldots & x_m(1) \\ x_1(2) & x_2(2) & \ldots & x_m(2) \\ \ldots & \ldots & \ldots & \ldots \\ x_1(N) & x_2(N) & \ldots & x_m(N) \end{bmatrix}, \ Y_N = \begin{bmatrix} y(1) \\ y(2) \\ \ldots \\ y(N) \end{bmatrix}$$

Coefficients of the regression equation:

$$A = \begin{bmatrix} a_1 \\ a_2 \\ \ldots \\ a_m \end{bmatrix}$$

Calculated (model) values of the output variable (actually, conditional mean values of the actual variable estimated using the regression model) are:

$$Y_N^{MOD} = \begin{bmatrix} y^{MOD}(1) \\ y^{MOD}(2) \\ \ldots \\ y^{MOD}(N) \end{bmatrix} = \begin{bmatrix} x_1(1) & x_2(1) & \ldots & x_m(1) \\ x_1(2) & x_2(2) & \ldots & x_m(2) \\ \ldots & \ldots & \ldots & \ldots \\ x_1(N) & x_2(N) & \ldots & x_m(N) \end{bmatrix} \cdot \begin{bmatrix} a_1 \\ a_2 \\ \ldots \\ a_m \end{bmatrix} = X_N \cdot A$$

Our goal is to minimize the accuracy criterion of the model, representing the "goodness" of the A values represented by a scalar Q(a) defined as follows,

$$Q(A) = \left(Y_N - Y_N^{MOD}\right)^T \left(Y_N - Y_N^{MOD}\right)$$

$$Q(A) = \left[Y_N^T - (X_N A)^T\right][Y_N - X_N A] = \left(Y_N^T - A^T X_N^T\right)(Y_N - X_N A)$$

$$Y_N^T Y_N - Y_N^T X_N A - A^T X_N^T Y_N + A^T X_N^T X_N A \rightarrow Min$$

It is known from the theory of statistical estimation that the minimum of the above expression is reached at $X_N^T X_N A = X_N^T Y_N$ thus the least square solution for coefficients A is:

$$A = \left(X_N^T X_N\right)^{-1} X_N^T Y_N$$

Case 2 could be reduced to Case 1 by introducing secondary variables $x_{m+1}(k) = x_1(k)^2$, $x_{m+2}(k) = x_2(k)^2$, ..., $x_{2m}(k) = x_m(k)^2$, $x_{2m+1}(k) = x_1(k)x_2(k)$, $x_{2m+2}(k) = x_1(k)x_3(k)$, ... While numerical values of the original (primary) variables $x_1(k)$, $x_2(k)$, ..., $x_m(k)$ are given, the secondary variables are to be calculated prior to forming matrix X_N. Note that vector A will be appropriately extended.

Case 3 could be reduced to Case 1 by introducing secondary variables $z_1(k) = \Phi_1[x_i(k),\ i = 1,2,\dots,m]$, $z_2(k) = \Phi_2[x_i(k),\ i = 1,2,\dots,m]$, $z_3(k) = \Phi_3[x_i(k),\ i = 1,2,\dots,m]$, ...

While numerical values of the original (primary) variables $x_1(k)$, $x_2(k)$, ..., $x_m(k)$ are given, the secondary variables are to be calculated prior to forming matrix Z_N similarly to X_N. Then vector A will be defined as $(Z_N^T Z_N)^{-1} Z_N^T Y_N$

Effect of Measurement Noise It is important to differentiate noise as a natural component of the input signal and noise caused by measurement/data recording errors. Being a component of the input signal ($v_1(k)$ and $v_2(k)$, see Fig. 2.1), noise

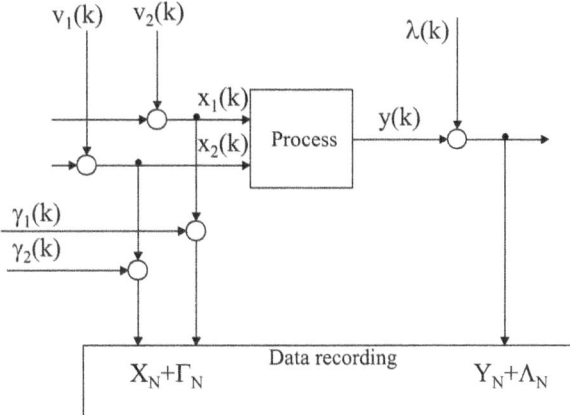

Fig. 2.1 Input and output measurement with noise

propagates through the system resulting in the additional variability of the output signal—this noise is known to be beneficial for parameter estimation. Measurement noise, $\gamma_1(k)$, $\gamma_2(k)$ and $\lambda(k)$, has a detrimental effect on parameter estimation. We will consider the measurement noise.

Assume that noise in the input channels, accompanying measurements X_N, is represented by matrix Γ_N and noise in the output, accompanying measurements Y_N, is represented by column Λ_N. The following characteristics of noise, quite realistic, are expected:

- noise is additive,
- noise is unbiased, i.e. has zero mean value,
- noise has a finite variance,
- noise is not correlated with input and output signals, $x_i(k)$, $i = 1,2,3,\ldots,m$, and $y(k)$, and
- there is no correlation between noise in particular input/output channels

Consider the effect of noise in the output on the parameter estimation process. Assume that in the original equation of the least-squares estimation matrix Y_N is replaced by $Y_N + \Lambda_N$, then

$$A = \left(X_N{}^T X_N\right)^{-1} X_N{}^T (Y_N + \Lambda_N) = \left(X_N{}^T X_N\right)^{-1} \left(X_N{}^T Y_N + X_N{}^T \Lambda_N\right)$$

Modify the above expression as follows,

$A = \left(\dfrac{1}{N} X_N{}^T X_N\right)^{-1} \left(\dfrac{1}{N} X_N{}^T Y_N + \dfrac{1}{N} X_N{}^T \Lambda_N\right)$. It could be seen that components

of column $-$ matrix $\dfrac{1}{N} X_N{}^T \Lambda_N$ are $\dfrac{1}{N} \sum\limits_{k=1}^{N} x_i(k)\lambda(k)$, where $i = 1, 2, \ldots, m$, i.e. are

covariance coefficients between input variables

and noise in the output, therefore as $\dfrac{1}{N} \sum\limits_{k=1}^{N} x_i(k)\lambda(k) \to 0$ as $N \to \infty$

Finally, for $N \gg 1$ $A = \left(X_N{}^T X_N\right)^{-1} X_N{}^T (Y_N + \Lambda_N) \approx \left(X_N{}^T X_N\right)^{-1} \left(X_N{}^T Y_N\right)$.

Therefore the following conclusion could be drawn: *noise in the output does not result in biased estimates of parameters A, however, an increased number of observations, N, should be processed to "average out" the effect of the noise.*

Consider the effect of noise in the input channels. Assume that in the original equation of the least-squares estimation matrix X_N is replaced by $X_N + \Gamma_N$, then

$$\left[(X_N + \Gamma_N)^T (X_N + \Gamma_N)\right]^{-1} (X_N + \Gamma_N)^T Y_N =$$

$$\left[\frac{1}{N} X_N{}^T X_N + \frac{1}{N} \Gamma_N{}^T X_N + \frac{1}{N} X_N{}^T \Gamma_N + \frac{1}{N} \Gamma_N{}^T \Gamma_N\right]^{-1} \left(\frac{1}{N} X_N{}^T Y_N + \frac{1}{N} \Gamma_N{}^T Y_N\right)$$

Then, due to the properties of the measurement noise

$$\lim_{N \to \infty} \left(\frac{1}{N} \Gamma_N^T X_N\right) = \lim_{N \to \infty} \left(\frac{1}{N} X_N^T \Gamma_N\right) = \lim_{N \to \infty} \left(\frac{1}{N} \Gamma_N^T Y_N\right) = 0$$

However $\lim_{N \to \infty} \left(\frac{1}{N} \Gamma_N^T \Gamma_N\right) = \begin{bmatrix} \sigma_1^2 & 0 & \cdots & 0 \\ 0 & \sigma_2^2 & \cdots & 0 \\ \cdots & \cdots & \cdots & \cdots \\ 0 & 0 & \cdots & \sigma_m^2 \end{bmatrix} \neq 0$, where σ_i^2, $i = 1, 2, \ldots,$

m, is the variance of noise in the ith input channel.

Therefore one has to conclude that *measurement noise in the input channels results in biased estimates of parameters* A, *regardless of the number of observations*, N.

Example 2.1 The purpose of this example is to illustrate the effect of measurement noise in the output on parameter estimation. Variables $x_1(i)$, $x_2(i)$, $x_3(i)$ and $x_4(i)$, $i = 1, 2, \ldots, 300$ were generated as combinations of several sinusoidal signals of different frequencies and amplitudes

$$x_1(i) = 3 * \sin(.2*i) + 1.1 * \sin(7*i) + .1 * \sin(7311*i)$$

$$x_2(i) = 5 * \sin(.01*i) + 2 * \sin(17*i) + .2 * \sin(17311*i) + .2 * x_1(i)$$

$$x_3(i) = 7 * \sin(.01*i) + .5 * \sin(3*i) + .05 * \sin(711*i) - .2 * x_1(i) \\ + .3 * x_2(i)$$

$$x_4(i) = \sin(.03*i) + .05 * \sin(13*i) + .01 * \sin(799*i) + .3 * x_1(i) \\ + .03 * x_2(i) + .07 * x_3(i)$$

and organized into array X_{300}. Variable y(i) is defined as

$$y(i) = 1 * x_1(i) - 2 * x_2(i) + 3 * x_3(i) + 4 * x_4(i) + \Delta(i), \ i = 1, .., 300$$

where $\Delta(i) = .1 * \sin(717*i)$ is a signal imitating the unbiased output noise, and organized in array Y_{300}.

Although "true" coefficients of the regression equation are known, $a_1 = 1$, $a_2 = -2$, $a_3 = 3$, $a_4 = 4$, let us attempt to estimate these coefficients using the first 50 rows of arrays X_{300} and Y_{300}, first 150 rows of arrays X_{300} and Y_{300}, and the entire arrays X_{300} and Y_{300}:

$$\left(X_{50}{}^T X_{50}\right)^{-1}\left(X_{50}{}^T Y_{50}\right) = \begin{bmatrix} 1.006 \\ -2.001 \\ 2.992 \\ 4.020 \end{bmatrix}$$

$$\left(X_{100}{}^T X_{100}\right)^{-1}\left(X_{100}{}^T Y_{100}\right) = \begin{bmatrix} 1.015 \\ -2.001 \\ 3.003 \\ 3.992 \end{bmatrix}$$

$$\left(X_{300}{}^T X_{300}\right)^{-1}\left(X_{300}{}^T Y_{300}\right) = \begin{bmatrix} 1.011 \\ -2.000 \\ 3.000 \\ 4.000 \end{bmatrix}$$

It could be seen that the output noise does result in the estimation error, however this error tends to decrease as the number of observations increases. It is good to note that the determinant of the covariance matrix of the input variables, $K_{XX} = 142.09 \neq 0$.

Example 2.2 Introduce unbiased "measurement noise" in the input channels of the previous problem,

$$n_1(i) = .4^* \sin(77^*i)$$
$$n_2(i) = .13^* \sin(177^*i)$$
$$n_3(i) = 3.1^* \sin(1771^*i)$$
$$n_4(i) = 1.1^* \sin(7177^*i), \ i = 1, 2, \ldots, 300$$

and organize it in array N_{300}. The output noise is removed, i.e.

$$y(i) = 1^*x_1(i) - 2^*x_2(i) + 3^*x_3(i) + 4^*x_4(i), \ i = 1, .., 300$$

Compute model parameters using array $\underline{X}_{300}$ "contaminated with noise", i.e. $\overline{X}_{300} = X_{300} + N_{300}$ and array Y_{300} free of noise:

$$\left(\overline{X}_{300}{}^T \overline{X}_{300}\right)^{-1}\left(\overline{X}_{300}{}^T Y_{300}\right) = \begin{bmatrix} .617 \\ 0.802 \\ 1.306 \\ 2.239 \end{bmatrix}$$

that is a completely wrong result.

Is it possible to improve this result using the information on measurement errors? Assume that the variances of the particular components of the measurement errors (constituting the input noise) are known: $\sigma_1^2 = 0.080$, $\sigma_2^2 = 0.008$, $\sigma_3^2 = 4.807$, and $\sigma_4^2 = 0.605$. Then the covariance matrix of noise could be constructed under the assumption that there is no cross-correlation between noise components:

$$K_{NN} = \begin{bmatrix} 0.080 & 0 & 0 & 0 \\ 0 & 0.008 & 0 & 0 \\ 0 & 0 & 4.807 & 0 \\ 0 & 0 & 0 & 0.605 \end{bmatrix}$$

Compute covariance matrices K_{XX} and K_{YX} using the available measurement data $\overline{X}_{300}$ and Y_{300}:

$$K_{XX} = \begin{bmatrix} 5.161 & 1.060 & -0.615 & 1.753 \\ 1.060 & 15.330 & 22.75 & 2.381 \\ -0.615 & 22.750 & 43.04 & 3.173 \\ 1.753 & 2.381 & 3.173 & 1.888 \end{bmatrix} \text{ and } K_{YX} = \begin{bmatrix} 7.155 \\ 48.000 \\ 81.200 \\ 11.370 \end{bmatrix}$$

Now the covariance matrix of the measurement noise, K_{NN}, can be utilized in the estimation task:

$$\left(\frac{1}{300} \overline{X}_{300}^T \overline{X}_{300} - K_{NN} \right)^{-1} \left(\frac{1}{300} \overline{X}_{300}^T Y_{300} \right) = (K_{XX} - K_{NN})^{-1} \left(\frac{1}{300} \overline{X}_{300}^T Y_{300} \right)$$

$$= \begin{bmatrix} 0.706 \\ -1.898 \\ 2.915 \\ 4.205 \end{bmatrix}$$

It could be seen that new estimates are not perfect, but are drastically better than the previous result. Why didn't we obtain a perfect result?—Recall that matrix K_{NN} was constructed under the assumption that there is no cross-correlation between noise components. Having the simulated "input noise array" N_{300}, we can obtain the "true" matrix K^{TRUE}_{NN} that is quite different from K_{NN}:

$$K^{TRUE}_{NN} = \frac{1}{300} N_{300}^T N_{300} = \begin{bmatrix} 0.080 & 1e-04 & 1e-04 & 0.219 \\ 1e-04 & 8e-03 & -2e-03 & 3e-04 \\ 1e-04 & -2e-03 & 4.807 & 1e-03 \\ 0.219 & 3e-04 & 1e-03 & 0.605 \end{bmatrix}$$

Effect of Cross-Correlation of Input Variables Imagine that two input variables, say x_i and x_j are highly correlated, to such extent that $x_i \approx c \cdot x_j + d$, where c and d are some constants. This condition will result in two rows of matrix $X_N^T X_N$

to be or almost be linearly dependent, therefore the determinant of this matrix Det $(X_N^T X_N) \approx 0$. This situation is known as a "poorly defined" parameter estimation problem. An "almost singular" matrix $X_N^T X_N$ presents a serious numerical difficulty in the calculation of coefficients A. In addition, one can see that since x_i duplicates x_j it does not make sense to introduce both variables in the mathematical model.

Principal Component Analysis is a numerical technique that addresses both issues.

Recall that

$$A = \left(X_N^T X_N\right)^{-1} X_N^T Y_N = \left(\frac{1}{N} X_N^T X_N\right)^{-1} \frac{1}{N} X_N^T Y_N = K_{XX}^{-1} K_{XY}$$

where K_{XX} and K_{YY} are the estimated covariance matrices of the appropriate variables. Assume that matrix K_{XX} is "almost singular", i.e. Det $(K_{XX}) \ll 1$.

Introduce a new set of input variables, $z_1(k), z_2(k), \ldots, z_m(k)$, obtained by a linear

transformation of the original variables $x_1(k), x_2(k), \ldots, x_m(k)$, i.e. $W \begin{bmatrix} z_1(k) \\ z_2(k) \\ \ldots \\ z_m(k) \end{bmatrix} =$

$\begin{bmatrix} x_1(k) \\ x_2(k) \\ \ldots \\ x_m(k) \end{bmatrix}$ or $\begin{bmatrix} z_1(k) \\ z_2(k) \\ \ldots \\ z_m(k) \end{bmatrix} = W^{-1} \begin{bmatrix} x_1(k) \\ x_2(k) \\ \ldots \\ x_m(k) \end{bmatrix}$ where W is a speciallydefined $(m \times m)$ matrix filter

Consequently, the regression equation (mathematical model) can be redefined as

$$y^{MOD}(k) = \sum_{j=1}^{m} b_j z_j(k).$$

Define matrix $Z_N = X_N W$, then coefficients b_j, $j = 1, 2, \ldots, m$, represented by vector B will be defined as

$$B = \left(Z_N^T Z_N\right)^{-1} Z_N^T Y_N = \left(\frac{1}{N} Z_N^T Z_N\right)^{-1} \frac{1}{N} Z_N^T Y_N = K_{ZZ}^{-1} K_{ZY}$$

where notations are self-explanatory. We want to assure that new variables, $z_1(k)$, $z_2(k), \ldots, z_m(k)$, are orthogonal, i.e. their covariance matrix K_{ZZ} is diagonal. This property of the new variables guarantees that the parameter estimation problem will be successfully solved.

Consider

$$K_{ZZ} = \frac{1}{N} Z_N^T Z_N = \frac{1}{N} (X_N W)^T (X_N W) = \frac{1}{N} W^T X_N^T X_N W = W^T \left(\frac{1}{N} X_N^T X_N\right)$$
$$W = W^T K_{XX} W$$

Written below is the matrix diagonalization formula (similarity transformation) known from matrix theory:

$$M^{-1}K_{XX}M = C,$$

where $C = \begin{bmatrix} \lambda_1 & 0 & \ldots & 0 \\ 0 & \lambda_2 & \ldots & 0 \\ \ldots & \ldots & \ldots & \ldots \\ 0 & 0 & \ldots & \lambda_m \end{bmatrix}$,

λ_j, $j = 1,2,\ldots m$, are eigenvalues of matrix K_{XX}, and

M is known as the *modal matrix* of matrix K_{XX}. It is good to remember that eigenvalues of matrix K_{XX} can be obtained by solving the following polynomial equation,

$$Det(\lambda I - K_{XX}) = 0$$

where I is a unity matrix, and all eigenvalues, λ_j, $j = 1,2,\ldots m$, of matrix $K_{XX} = \dfrac{1}{N}X_N{}^T X_N$ are real and positive.

Recall now that a modal matrix could be defined for any square matrix that has real eigenvalues, say K_{XX}. If matrix K_{XX} has the dimension of $m \times m$, then its modal matrix is also a $m \times m$ matrix. A modal matrix is formed as follows: its first column is the eigenvector of matrix K_{XX} corresponding to the first eigenvalue λ_1, its second column is the eigenvector of matrix K_{XX} corresponding to the second eigenvalue λ_2, and so on. Finally, an eigenvector of a square matrix, say K_{XX} corresponding to its eigenvalue λ_j is defined by

- forming matrix $\lambda_j I - K_{XX}$ where I is a unit matrix
- taking the adjoint of this matrix $D = Adj(\lambda_j I - K_{XX})$
- using any column of matrix D as the eigenvector of matrix K_{XX} corresponding to its eigenvalue λ_j.

Note that the matrix diagonalization is a computationally intensive procedure that is routinely performed by engineering software tools. Keep in mind that any constant multiplied by an eigenvector is still the same eigenvector, therefore many software tools generate apparently different sets of eigenvectors for the same matrix. The following numerical example features a matrix X_N, matrix K_{XX}, its eigenvalues and eigenvectors, its modal matrix M, and the diagonalization of matrix K_{XX}

Example 2.3 Definition and diagonalization of a covariance matrix.

$$
X_{10} = \begin{bmatrix} 1 & 4 & 4 & 4 \\ 2 & 8 & 6 & -4 \\ 3 & -7 & 3 & 2 \\ 1 & 5 & 2 & 3 \\ 5 & 2 & 1 & -2 \\ 7 & 1 & 1 & 1 \\ 9 & -1 & 4 & -3 \\ -6 & -6 & -4 & 5 \\ -1 & 7 & 9 & -5 \\ 2 & 0 & -7 & 8 \end{bmatrix}, \quad K_{XX} = \frac{1}{10} X_{10}{}^{T} X_{10} = \begin{bmatrix} 21.1 & 4.1 & 7.6 & -3.4 \\ 4.1 & 24.5 & 13.9 & -8 \\ 7.6 & 13.9 & 22.9 & -13 \\ -3.4 & -8 & -13 & 17.3 \end{bmatrix},
$$

$$
\begin{bmatrix} \lambda_1 \\ \lambda_2 \\ \lambda_3 \\ \lambda_4 \end{bmatrix} = \begin{bmatrix} 5.756 \\ 12.961 \\ 18.664 \\ 48.419 \end{bmatrix}
$$

$$
M = \begin{bmatrix} -.163 & .156 & .922 & .313 \\ -.251 & .702 & -.355 & .564 \\ .736 & -.258 & -.038 & .625 \\ .607 & .645 & .147 & -.440 \end{bmatrix} \quad \text{It could be seen that}
$$

$$
\begin{bmatrix} -.163 & .156 & .922 & .313 \\ -.251 & .702 & -.355 & .564 \\ .736 & -.258 & -.038 & .625 \\ .607 & .645 & .147 & -.440 \end{bmatrix}^{-1} \times \begin{bmatrix} 21.1 & 4.1 & 7.6 & -3.4 \\ 4.1 & 24.5 & 13.9 & -8 \\ 7.6 & 13.9 & 22.9 & -13 \\ -3.4 & -8 & -13 & 17.3 \end{bmatrix}
$$

$$
\times \begin{bmatrix} -.163 & .156 & .922 & .313 \\ -.251 & .702 & -.355 & .564 \\ .736 & -.258 & -.038 & .625 \\ .607 & .645 & .147 & -.440 \end{bmatrix} = \begin{bmatrix} 5.756 & 0 & 0 & 0 \\ 0 & 12.961 & 0 & 0 \\ 0 & 0 & 18.664 & 0 \\ 0 & 0 & 0 & 48.419 \end{bmatrix}
$$

Now one additional piece of information: it is known from the matrix theory that if a matrix is defined as $K_{XX} = \frac{1}{N} X_N{}^{T} X_N$ then the transpose of its modal matrix is equivalent to the inverse. Let us demonstrate this property using our example:

$$
\begin{bmatrix} -.163 & .156 & .922 & .313 \\ -.251 & .702 & -.355 & .564 \\ .736 & -.258 & -.038 & .625 \\ .607 & .645 & .147 & -.440 \end{bmatrix}^{-1} = \begin{bmatrix} -.163 & -.251 & .736 & .607 \\ .156 & .702 & -.258 & .645 \\ .922 & -.355 & -.038 & .147 \\ .313 & .564 & .625 & -.440 \end{bmatrix}
$$

$$
= \begin{bmatrix} -.163 & .156 & .922 & .313 \\ -.251 & .702 & -.355 & .564 \\ .736 & -.258 & -.038 & .625 \\ .607 & .645 & .147 & -.440 \end{bmatrix}^{T}
$$

Now one can see that matrix filter W, utilized to convert "old" input variables, $x_j(k)$, $j = 1,2,\ldots$, m, into the new orthogonal variables, $z_j(k)$, $j = 1,2,\ldots,$m, is nothing but the modal matrix M of the covariance matrix K_{XX}.

Now define the array of diagonalized variables $Z_{10} = X_{10}M$:

$$Z_{10} = \begin{bmatrix}
4.204 & 4.515 & -0.061 & 3.307 \\
-0.346 & 1.803 & -1.813 & 10.646 \\
4.692 & -3.928 & 5.431 & -2.014 \\
1.874 & 5.087 & -0.487 & 3.062 \\
-1.795 & 0.638 & 3.569 & 4.199 \\
-0.049 & 2.184 & 6.211 & 2.941 \\
-0.092 & -2.260 & 8.062 & 6.074 \\
2.575 & -0.895 & -2.515 & -9.962 \\
1.995 & -0.786 & -4.488 & 11.457 \\
-0.624 & 7.277 & 3.291 & -7.267
\end{bmatrix}$$

and check their covariance matrix

$$K_{ZZ} = \frac{1}{10} Z_{10}{}^T Z_{10} = \begin{bmatrix}
5.756 & 0 & 0 & 0 \\
0 & 12.961 & 0 & 0 \\
0 & 0 & 18.664 & 0 \\
0 & 0 & 0 & 48.419
\end{bmatrix}$$

Note a useful fact: the eigenvalues of the covariance matrix K_{XX}, λ_k, are equal to the "variance + mean value squared" of the respective component of the diagonalized vector, z_k.

Application of the Principal Component Analysis (PCA) First, let us modify the modeling problem: the model to be established is

$$y^{MOD}(i) = \sum_{k=1}^{m} a_k x_k(i) = \sum_{k=1}^{m} b_k z_k(i)$$

where $z_k(i)$, $k = 1,2,\ldots,$m are orthogonalized variables. It could be seen that, in principle, the parameter estimation task implies the solution of the familiar equation,

$$B = (Z_N{}^T Z_N)^{-1}(Z_N{}^T Y_N) = \left(\frac{1}{N} Z_N{}^T Z_N\right)^{-1} \left(\frac{1}{N} Z_N{}^T Y_N\right) = K_{ZZ}{}^{-1} K_{YZ}$$

where K_{ZZ} and K_{YZ} are appropriate covariance matrices.

When parameters B are known, it is easy to convert them into parameters A of the original regression equation of interest, indeed $Y_N{}^{MOD} = X_N \cdot A = Z_N \cdot B = X_N \cdot M \cdot B$, therefore, $A = M \cdot B$

The important part of the PCA is the fact that estimation of coefficients B does not lead to a system of simultaneous linear equations and does not require matrix inversion, indeed,

$$B = K_{ZZ}^{-1}K_{ZY} = \begin{bmatrix} \lambda_1 & 0 & \cdots & 0 \\ 0 & \lambda_2 & \cdots & 0 \\ \cdots & \cdots & \cdots & \cdots \\ 0 & 0 & \cdots & \lambda_m \end{bmatrix}^{-1}.$$

$$K_{ZY} = \begin{bmatrix} \dfrac{1}{\lambda_1} & 0 & \cdots & 0 \\ 0 & \dfrac{1}{\lambda_2} & \cdots & 0 \\ \cdots & \cdots & \cdots & \cdots \\ 0 & 0 & \cdots & \dfrac{1}{\lambda_m} \end{bmatrix} \cdot \begin{bmatrix} K_{z_1y} \\ K_{z_2y} \\ \cdots \\ K_{z_my} \end{bmatrix} = \begin{bmatrix} \dfrac{K_{z_1y}}{\lambda_1} \\ \dfrac{K_{z_2y}}{\lambda_2} \\ \cdots \\ \dfrac{K_{z_my}}{\lambda_m} \end{bmatrix}$$

i.e. every coefficient b_k, $k = 1,2,\ldots,m$, can be calculated using simple scalar expressions.

Imagine that the computation of coefficients of a regression equation is part of an on-going control procedure. Instead of utilizing the original matrix equation of the LSM and risking possible computational instability due to "almost singular" matrix K_{XX}, it makes sense to obtain orthogonalized variables, estimate coefficients B, avoiding matrix inversion, and then convert them into coefficient A.

It is important to realize that variables z_j, $j = 1,2,\ldots, m$, are mutually orthogonal and therefore do not duplicate each other. Knowing the individual contributions of these variables into the model would allow excluding those least contributive thus simplifying the model. Then how can these individual contributions be evaluated?—Variability of the model is defined by the variance

$$\sigma_{MOD}^2 = \frac{1}{N}\sum_{i=1}^{N}\left[Y^{MOD}(i) - \overline{Y}^{MOD}\right]^2 = \frac{1}{N}\sum_{i=1}^{N}\left[\sum_{k=1}^{m}b_kz_k(i) - \overline{Y}^{MOD}\right]^2$$

$$= \frac{1}{N}\sum_{i=1}^{N}\left[\sum_{k=1}^{m}b_k[z_k(i) - \overline{z}_k]\right]^2$$

where $\overline{Y}^{MOD}$ and $\overline{z}_k$, $k = 1,2,\ldots,m$, are mean values of the respective variables. Reversing the order of summation in the above expression and assuming that $N \gg 1$ results in

$$\sigma_{MOD}^2 = \sum_{k=1}^{m}\left[\frac{1}{N}b_k^2\sum_{i=1}^{n}[z_k(i) - \overline{z}_k]^2\right] = \sum_{k=1}^{m}b_k^2\sigma_k^2$$

where σ_k^2 is the variance of the kth orthogonalized variable.

At this point, the only unknowns are variances σ_k^2, $k = 1,2,\ldots,m$. They can be found by computing the mean values of these variables, $\bar{z}_k$, $k = 1,2,\ldots,m$, and then computing every $\sigma_k^2 = \lambda_k - \bar{z}_k^2$ $k = 1,2,\ldots,m$.

Finally, individual contributions (in percent) of particular variables z_k into the regression model can be expressed as

$$\delta_j = \frac{b_j^2 \sigma_j^2}{\sum\limits_{k=1}^{m} b_k^2 \sigma_k^2}, \quad j = 1, 2, \ldots, m$$

Therefore, PCA provides an opportunity for choosing a subset of orthogonalized variables z_j whose combined contribution to the regression model is sufficiently high, say 90%, and discarding the rest of these variables as non-contributive.

Assume that according to the above analysis, variable z_H is the least contributive component of the vector of orthogonalized variables. Then the column #H of the array Z_N can be discarded resulting in array $\hat{Z}_N$ and parameters $\hat{B}$ will be defined as follows:

$$\hat{B} = \left(\hat{Z}_N^T \hat{Z}_N\right)^{-1}\left(\hat{Z}_N^T Y_N\right)$$

Note that vector $\hat{B}$ has one component less than vector $B = \left(Z_N^T Z_N\right)^{-1}\left(Z_N^T Y_N\right)$, i.e. $\hat{B}$ is a $m-1$ vector. Remove column #H from the modal matrix M that will result in matrix $\hat{M}$ that has m column and $m-1$ rows. Finally, vector of the model parameters A will me defined as

$$A = \hat{M}\hat{B}$$

Example 2.4 Application of the Principal Component Analysis, a simulation case study

The following are the definitions of variables $x_k(i)$, $k = 1,2,3,4$, and $y(i)$:

$$x_1(i) = 3* \sin(.01*i) + .5* \sin(7*i) + .1* \sin(177*i)$$
$$x_2(i) = 6* \sin(.031*i) + .2* \sin(17*i) + .3* \sin(7177*i)$$
$$x_3(i) = 4* \sin(.077*i) + .2* \sin(74*i) + .2* \sin(7111*i)$$
$$x_4(i) = x_3(i) + .00002*\sin(17171*i)$$
$$y(i) = 3* x_1(i) - 2*x_2(i) + 5* x_3(i) - 3* x_4(i) + .07* \sin(817371*i),$$
$$i = 1, 2, 3, \ldots, 500$$

This choice of variables reflects our intention to create a case when the first three input variables are virtually independent, but the fourth input variable, x_4, is almost identical (highly correlated) to x_3. In addition, it could be expected that mean values of the input variables be close to zero. According to the definition of output variable $y(i)$, the "true" values of the appropriate regression coefficients are 3, -2, 5, -3. However, this relationship is quite superficial: since $x_4(i) \approx x_3(i)$, the relationship is effectively $y(i) = 3^* x_1(i) - 2^* x_2(i) + 5^* x_3(i) - 3^* x_3(i) = 3^* x_1(i) - 2^* x_2(i) + 2^* x_3(i)$, however this reality is not apparent to the individual who attempts to establish the regression model relating $y(i)$ to $x_1(i)$, $x_2(i)$, $x_3(i)$, $x_4(i)$.

First, the above variables are to be organized into arrays X_{500} and Y_{500}, then an attempt to solve the LSM problem as follows could be made:

$$\left(X_{500}{}^T X_{500}\right)^{-1}\left(X_{500}{}^T Y_{500}\right) = \begin{bmatrix} 3.000 \\ -2.000 \\ 1.050 \\ 0.950 \end{bmatrix}$$

We do know that this result is erroneous, but there is a good reason to doubt this result even if one does not know the "true" coefficient values, the covariance matrix K_{XX} is "almost singular"

$$\mathrm{Det}(K_{XX}) = \mathrm{Det} \begin{bmatrix} 4.890 & -1.197 & 0.209 & 0.209 \\ -1.197 & 18.30 & -0.313 & -0.313 \\ 0.209 & -0.313 & 7.943 & 7.943 \\ 0.209 & -0.313 & 7.943 & 7.943 \end{bmatrix} = 1e - 07$$

Consequently, this situation presents an ideal target for PCA.
Obtain the modal matrix of K_{XX} and its eigenvalues:

$$M = \begin{bmatrix} -0.090 & 7e-03 & 0.996 & -1e-09 \\ 0.980 & 0.178 & 0.087 & 7e-10 \\ -0.125 & 0.696 & -0.016 & -0.707 \\ -0.125 & 0.696 & -0.016 & 0.707 \end{bmatrix} \text{ and the eigenvalues are } \begin{bmatrix} \lambda_1 \\ \lambda_2 \\ \lambda_3 \\ \lambda_4 \end{bmatrix}$$

$$= \begin{bmatrix} 18.49 \\ 15.81 \\ 4.778 \\ 1e-10 \end{bmatrix}$$

It is known that eigenvalues represent the variability of the particular components of the vector of orthogonalized variables whose values are defined by the array $Z_{500} = X_{500}M$, indeed

$$\frac{1}{500} Z_{500}{}^T Z_{500} = \begin{bmatrix} 18.49 & -1e-15 & 2e-15 & -1e-15 \\ -1e-15 & 15.81 & 8e-16 & -2e-15 \\ 2e-15 & 8e-16 & 4.778 & -6e-17 \\ -1e-15 & -2e-15 & -6e-17 & 1e-10 \end{bmatrix}$$

This indicates that the fourth component of array Z_{500} is irrelevant and the first three columns of the array Z_{500} are sufficient for the modeling process. With the trimmed array $\hat{Z}_{500}$ the estimation results are as follows:

$$B = \left(\hat{Z}_{500}^{\mathrm{T}}\hat{Z}_{500}\right)^{-1}\left(\hat{Z}_{500}^{\mathrm{T}}Y_{500}\right) = \begin{bmatrix} -2.481 \\ 1.056 \\ 2.780 \end{bmatrix}$$

This result could be trusted:

$$K_{ZZ} = \frac{1}{500}\hat{Z}_{500}^{\mathrm{T}}\hat{Z}_{500} = \begin{bmatrix} 18.49 & -1e-15 & 2e-15 \\ -1e-15 & 15.81 & 8e-16 \\ 2e-15 & 8e-16 & 4.778 \end{bmatrix} \text{ and } \mathrm{Det}(K_{ZZ}) \approx 1e+03$$

Finally, convert coefficients B into the required coefficients A. Note that since the last column of array Z_{500} was eliminated, the last column of matrix M must be also crossed out:

$$\hat{A} = \hat{M}B = \begin{bmatrix} -0.090 & 7e-03 & 0.996 \\ 0.980 & 0.178 & 0.087 \\ -0.125 & 0.696 & -0.016 \\ -0.125 & 0.696 & -0.016 \end{bmatrix} \begin{bmatrix} -2.481 \\ 1.056 \\ 2.780 \end{bmatrix} = \begin{bmatrix} 3.000 \\ -2.000 \\ 1.000 \\ 1.000 \end{bmatrix}$$

Note that since $x_3 \approx x_4$, this result is completely consistent with the way the original data was generated.

2.2 Validation of Regression Models

While the computational task of the least squares method is very straight forward, one should realize that

- the obtained regression equation may or may not be statistically valid, i.e. it may not reflect the existing trend-type relationship, but just "numerical noise",
- obtained parameter values are only statistical estimates and the "true" values could be found within some confidence intervals,
- computation of $y^{\mathrm{MOD}}(k)$ for any combination of input variables, $x_j(k)$, $j = 1,2,...,m$, results only in a statistical estimate, and the "true" value could be found within a confidence interval.

Let us address these issues.

Coefficient of Determination Recall that regression-based models do not represent a functional relationship between variable $y(k)$ and inputs $x_j(k)$, $j = 1,2,...,m$ but rather a trend. Indeed, variable $y(k)$ depends on a very large number of random factors, and only partially depends on the chosen inputs (regressors). It is very important to determine to what extent variable y(k) depends on the inputs of the model and to what extent it depends on the factors not included in the model.

It is known that all factors, included and not included in the model result in variability of the measured (true) value of variable y(k). This variability is represented by the *natural variance* of y(k) defined as

$$\sigma_y^2 = \frac{1}{N-1} \sum_{k=1}^{N} \left(y(k) - M_y \right)^2$$

where M_y is the estimated mean value of y(k). When parameters of the model a_j, $j = 1, 2, \ldots, m$, are established, variations of the input variables of the regression equation, x_j, $j = 1, 2, \ldots, m$, result in the variability of the output variable of the regression equation, $y^{MOD}(k)$ that is characterized by the model variance,

$$\sigma_{MOD}^2 = \frac{1}{N-1} \sum_{k=1}^{N} \left(y^{MOD}(k) - M_y \right)^2$$

(Note that due to one of the known properties of the LSM mean values of y(k) and $y^{MOD}(k)$ are the same). Finally, the variance of the model error, $e(k) = y(k) - y^{MOD}(k)$, that could be easily defined as

$$\sigma_E^2 = \frac{1}{N-1} (Y_N - X_N \cdot A)^T (Y_N - X_N \cdot A),$$

represents the variability of y(k) caused by all factors not included in the regression equation.

It is known that $\sigma_y^2 = \sigma_{MOD}^2 + \sigma_E^2$. The coefficient of determination represents the ratio between the joint effect of the factors included in the regression equation and the joint effect of all factors (included and not included) on the variable y(k), it is therefore defined as

$$\eta = \frac{\sigma_{MOD}^2}{\sigma_y^2} = \frac{\sigma_y^2 - \sigma_E^2}{\sigma_y^2} = 1 - \frac{\sigma_E^2}{\sigma_y^2}.$$

Coefficient of determination is always positive and does not exceed 1, i.e. $0 \leq \eta \leq 1$. It can be seen that it presents a "measure of goodness" of a regression equation, approaching 1 for very accurate models. Traditionally, for a successfully established regression equation, the coefficient of determination is expected to be at least 0.85.

It is important that coefficient of determination can be used for both linear and nonlinear equations. In the case of a linear regression equation, the coefficient of determination is equal to the multiple correlation coefficient squared.

Statistical Validation of a Regression Equation Coefficient of determination provides a rather qualitative way of statistical validation: a regression model is definitely valid if $\eta \geq 0.85$ and is definitely invalid if $\eta \leq .3$, however there is a more formal way to claim statistical significance or insignificance of a model. It is also based on comparing variances σ_y^2, σ_{MOD}^2, and σ_E^2 utilizing the Fisher distribution. However, it is important to recognize that the modeling techniques presented in this course are intended for the analysis and design of control systems, not for the "better understanding of complex laws in nature and society." Regression equations developed herein will be eventually reformulated as transfer functions. Consequently, the only mathematical models that are of any use in our application, are those "definitely valid." Therefore, we will use only one criterion of the validity of a model: its coefficient of determination must be at least 0.85. This would eliminate the use of the Fisher criterion, which allows one to justify the validity of a model that correctly reflects only 50 % of the variability of an output variable.

For example, let "true" relationship, $y(i) = 3x_1(i) - 2x_2(i) + 5x_3(i) - 3x_4(i)$, be modeled by a regression equation $y^{MOD}(i) = 2.989x_1(i)$ obtained using $y(i)$ and $x_1(i)$ observations. According to Fisher criterion this model could easily be statistically valid, because it *correctly* describes the existing relationship between variables $y(i)$ and $x_1(i)$. However, such a model is absolutely meaningless for the control process when in order to maintain the desired value of the output *all relevant* variables must be taken into account. Indeed, requesting the coefficient of determination to be at least 0.85 is a good way to qualify a mathematical model to be used as a transfer function.

Example 2.5 Application of the coefficient of determination for model simplification is illustrated by the following table. Assume that the available measurements of the output and input variables are assembled into eight various configurations of array X_N. The following table contains regression models obtained by solving the matrix equation of the LSM:

#	Regression equation	σ_y^2	σ_E^2	η	Validity
1	$y^{MOD} = 3.5x_1 + 4.2x_2 + .5x_1^2 + 1.1x_2^2 - 1.3x_1x_2$	12.4	.86	.931	Valid
2	$y^{MOD} = 3.5x_1 + 4.2x_2 + .5x_1^2 + 1.1x_2^2$	12.4	1.28	.897	Valid
3	$y^{MOD} = 3.5x_1 + 4.2x_2 + .5x_1^2 - 1.3x_1x_2$	12.4	1.36	.89	Valid
4	$y^{MOD} = 3.5x_1 + 4.2x_2 + 1.1x_2^2 - 1.3x_1x_2$	12.4	1.12	.91	Valid
5	$y^{MOD} = 3.5x_1 + .5x_1^2 + 1.1x_2^2 - 1.3x_1x_2$	12.4	4.39	.646	Invalid
6	$y^{MOD} = 4.2x_2 + .5x_1^2 + 1.1x_2^2 - 1.3x_1x_2$	12.4	5.58	.55	Invalid
7	$y^{MOD} = 3.5x_1 + 4.2x_2 + .5x_1^2$	12.4	1.52	.877	Valid
8	$y^{MOD} = 3.5x_1 + 4.2x_2 - 1.3x_1x_2$	12.4	1.41	.886	Valid
9	$y^{MOD} = 3.5x_1 + 4.2x_2$	12.4	2.04	.835	Invalid

It can be seen that the applied procedure includes estimation of the "natural" variance of variable y, estimation of the regression coefficients for particular combinations of column of the array X_N, computation of the standard deviation of the modeling error, computation of the coefficient of determination, and finally making a conclusion on the validity of the model. It is seen that cases 7 and 8 represent the simplest valid models.

Confidence Intervals of the Model Parameters Due to limited number of observations, each model parameter estimated by the least squares procedure is a random variable. Assume that a_j $(j = 1,2,\ldots,m)$ is one of the parameters estimated by the least squares procedure, and a_j^{TRUE} is the "true" value of this parameter that could be obtained by processing infinite number of observations, or by averaging all possible values of estimates a_j. Then it is known that with probability $P = 1 - 2\alpha$

$$| a_j - a_j^{TRUE} | \leq t(\alpha, n) S_E Q_{jj}^{1/2}$$

where
$j = 1,2,\ldots,m$ is the parameter index,
$t(\alpha,n)$ is t-distribution for the significance level α, and number of degrees of
    freedom $n = N$-m,
Q_{jj} is the jth diagonal element of the matrix $(X_N^T X_N)^{-1}$, and
S_E is standard deviation of the modeling error.

Note that for $N \gg 1$, $X_N^T X_N \approx N \cdot K_{XX}$ where K_{XX} is the covariance matrix of the input vector. Then the above definition of the width of the confidence interval could be rewritten as

$$\left| a_j - a_j^{TRUE} \right| = \frac{t(a, N - m) \cdot S_E \cdot \sqrt{q_{jj}}}{\sqrt{N}}$$

where q_{jj} is the jj − th diagonal element of matrix K_{XX}^{-1}

Confidence Intervals for Model-Based Prediction of the Output Variable Due to the stochastic nature of the relationship between input variables $x_j(k)$, $j = 1,2,\ldots,m$, and variable $y(k)$, the stochastic nature of estimated model parameters, a_j, $j = 1,2,\ldots,m$, and possible measurement noise, the calculated value of the output variable, $y^{MOD}(k)$, and its measured value, $y(k)$ are expected to be different. However, it is known that with probability $P = 1 - 2\alpha$

$$| y(k) - y^{MOD}(k) | \leq t(\alpha, n) S_E [x(k)^T Q\, x(k)]^{1/2}$$

where
$x(k) = [x_1(k)\ x_2(k) \ldots x_m(k)]^T$ is the vector of the input variables,
$Q = (X_N^T X_N)^{-1}$
$y^{MOD}(k) = \sum_{j=1}^{m} a_j x_j(k) = x^T(k) A$ and $y(k)$ are the model and measured values of
the output variable, respectively.

Again for $N \gg 1$ the width of the confidence interval could be redefined as

$$\left| y(k) - y^{MOD}(k) \right| = \frac{t(\alpha, N - m) \cdot S_E \cdot \sqrt{x(k)^T \cdot K_{XX}^{-1} \cdot x(k)}}{\sqrt{N}}$$

Example 2.6 The model $y = 4.969x_1 - 4.012\,x_2 + 2.65x_3 - 2.093x_4 + 2.063x_5$ was obtained using 300 observations of $y(k)$, $x_1(k)$, $x_2(k)$, $x_3(k)$, $x_4(k)$, and $x_5(k)$. Variance of the modeling error $S_E^2 = 29.49 = 5.43^2$. Obtain 95% confidence intervals for model parameters.

First, find the covariance matrix $X_{300}{}^T X_{300}$ and its inverse:

$$K_{XX} = \begin{bmatrix} 708.0 & 253.6 & 347.9 & -130.1 & 103.0 \\ 253.6 & 3e+03 & 1e+03 & 31.32 & 432.5 \\ 347.9 & 1e+03 & 2e+03 & 0.631 & 463.1 \\ -130.1 & 31.32 & 0.631 & 949.6 & 79.94 \\ 103.0 & 432.5 & 463.1 & 79.94 & 170.0 \end{bmatrix}$$

$K_{XX}{}^{-1}$
$$= \begin{bmatrix} 1.645209e-003 & 4.118080e-005 & -1.141253e-004 & 3.026922e-004 & -9.330502e-004 \\ 4.118080e-005 & 6.112089e-004 & 2.546555e-005 & 1.294520e-004 & -1.710311e-003 \\ -1.141253e-004 & 2.546555e-005 & 2.855153e-003 & 6.623047e-004 & -8.085377e-003 \\ 3.026922e-004 & 1.294520e-004 & 6.623047e-004 & 1.337917e-003 & -2.946240e-003 \\ -9.330502e-004 & -1.710311e-003 & -8.085377e-003 & -2.946240e-003 & 0.0342119 \end{bmatrix}$$

Finding the t-distribution value, $t(.025, 300) = t(.025, \infty) = 1.96$, and the following "widths" of the confidence intervals:

$$\Delta_1 = 1.96 \times 5.43 \times (1.645e-3)^{1/2} = 0.431$$
$$\Delta_2 = 1.96 \times 5.43 \times (6.112e-4)^{1/2} = 0.263$$
$$\Delta_3 = 1.96 \times 5.43 \times (2.855e-3)^{1/2} = 0.569$$
$$\Delta_4 = 1.96 \times 5.43 \times (1.338e-3)^{1/2} = 0.389$$
$$\Delta_5 = 1.96 \times 5.43 \times (0.0342)^{1/2} = 1.968$$

Then, with 95 % probability "true" parameter values will stay within the following limits:

$$4.538 \leq a_1{}^{TRUE} \leq 5.4,$$
$$-4.275 \leq a_2{}^{TRUE} \leq -3.749,$$
$$2.081 \leq a_3{}^{TRUE} \leq 3.219,$$

$$-2.482 \leq a_4{}^{\text{TRUE}} \leq -1.704,$$

$$0.095 \leq a_4{}^{\text{TRUE}} \leq 4.031$$

We have the "luxury" of knowing the "true" parameter values, 5, −4, 3, −2, and 1 and can conclude that they are indeed within the confidence intervals.

Example 2.7 Given a set of particular values of the input variables, $x_1(k) = 1.0$, $x_2(k) = 4.0$, $x_3(k) = 10.0$, $x_4(k) = 3.5$, and $x_5 = 12.3$. Estimate the value of the output variable $y(k)$ using the model,

$$y = 4.969x_1 - 4.012x_2 + 2.65x_3 - 2.093x_4 + 2.063x_5$$

and obtain a 95% confidence interval for the estimate.

$$\begin{aligned} Y^{\text{MOD}}(k) &= 4.969 \times 1.0 - 4.012 \times 4.0 + 2.65 \times 10.0 - 2.093 \times 3.5 + 2.063 \\ &\quad \times 12.3 \\ &= 33.469 \end{aligned}$$

To determine the confidence interval, note that

$t(.025, \infty) = 1.96$, $S_E = 5.43$,
$[x(k)^T Q x(k)]^{1/2} = 1.763$
"Width" of the confidence interval: $\Delta = 1.96 \times 5.43 \times 1.763 = 18.76$, therefore, $33.469 - 18.76 \leq y(k) \leq 33.469 + 18.76$ and, finally, $P\{14.709 \leq y^{\text{TRUE}}(k) \leq 52.229\} = 95\%$

Exercise 2.1

Generate input and the output variables as follows ($k = 1, 2, \ldots, 500$):

$$x_1(k) = 5 + 3 \cdot \text{Sin}(17 \cdot k) + \text{Sin}(177 \cdot k) + .3 \cdot \text{Sin}(1771 \cdot k)$$
$$x_2(k) = 1 - 2 \cdot \text{Sin}(91 \cdot k) + \text{Sin}(191 \cdot k) + .2 \cdot \text{Sin}(999 \cdot k)$$

$$x_3(k) = 3 + \text{Sin}(27 \cdot k) + .5 \cdot \text{Sin}(477 \cdot k) + .1 \cdot \text{Sin}(6771 \cdot k)$$

$$x_4(k) = -.1 \cdot x_1(k) + 3 \cdot x_2(k) + .5 \cdot \text{Sin}(9871 \cdot k) + .7 \cdot \text{Cos}(6711 \cdot k)$$

$$y(k) = 2 \cdot x_1(k) + 3 \cdot x_2(k) - 2 \cdot x_3(k) + 5 \cdot x_4(k) + .3 \cdot \text{Sin}(1577 \cdot k)$$
$$+ .2 \cdot \text{Cos}(7671 \cdot k)$$

Problem 1 Obtain "unknown" coefficients of the regression equation using the least squares method using the first 30 rows of arrays X_{500}, Y_{500}, first 100 rows, first 200 rows, and finally all 500 rows. Compare coefficients with the "true" coefficients and discuss your conclusions.

Problem 2 Compute 95 % confidence intervals for the coefficients of the model based on 500 observations. Check if the "true" coefficients are within the confidence intervals.

Problem 3 Assume $x(k) = [2.5, 3.0, -6.3, 10.]^T$ and compute $y(k)$ using the "true" relationship between the variables. Compute $y^{MOD}(k)$ using the final model and compute its 95 % confidence interval. Check if the "true" value, $y(k)$, is within the confidence intervals.

Problem 4 Given covariance matrices

$$K_{XX} = \begin{bmatrix} 3.084 & 0.846 & 1.158 & -0.434 & 0.343 \\ 0.846 & 10.90 & 4.012 & 0.104 & 1.442 \\ 1.158 & 4.012 & 6.269 & 1e-03 & 1.543 \\ -0.434 & 0.104 & 1e-03 & 3.425 & 0.266 \\ 0.343 & 1.442 & 1.543 & 0.266 & 0.677 \end{bmatrix} \quad \text{and} \quad K_{YX} = \begin{bmatrix} 12.970 \\ -19.470 \\ 7.213 \\ -8.649 \\ 0.610 \end{bmatrix}$$

estimated using measurement data. It is known that measurement errors in the corresponding input channels have the following variances: $\sigma_1^2 = 0.75$, $\sigma_2^2 = 1.66$, $\sigma_3^2 = 0.96$, $\sigma_4^2 = .26$, $\sigma_5^2 = 0.11$. Compute parameters of the mathematical model using this information. Compute parameter estimation errors caused by input noise.

Problem 5 Redefine variables $x_4(k)$ and $y(k)$ as follows:

$$x_4(k) = x_3(k) + .000012 \cdot Cos(7671 \cdot k),$$

$$y(k) = 2 \cdot x_1(k) + 3 \cdot x_2(k) - 2 \cdot x_3(k) + 5 \cdot x_4(k) + .3 \cdot Sin(1577 \cdot k) + .2$$
$$\cdot Cos(7671 \cdot k), \quad k = 1, 2, \ldots, 500$$

Apply principal component analysis for the evaluation of the regression parameters.

2.3   Recursive Parameter Estimation

Assume that input/output measurement data is given by arrays: X_N and Y_N and the model configuration is defined. Assume that model parameters are calculated via the Least Squares Method (LSM):

$$A_N = \left[X_N^T X_N\right]^{-1} \left[X_N^T Y_N\right]$$

How can an additional observation, $y(N+1)$, $x(N+1) = [x_1(N+1), \ldots, x_m(N+1)]$ be utilized?

But, should this additional measurement be utilized in the first place?—Yes, it should. First, because real measurement data is contaminated by noise and the estimation accuracy improves with the increased amount of data. Second, because the process described by input/output measurements may exhibit slow changes that must be reflected by its model by using the most recent measurements. Recursive Least Squares Method (RLSM) theory provides the necessary mechanism for the incorporation of additional data in already existing model, thus continuous model updating utilizing the flow of incoming measurements.

The RLSM is consistent with the principle of recursive estimation mentioned in Chap. 1. It does not require the accumulation of measurement arrays X_N and Y_N and consequent processing of these arrays. Instead, it allows processing one observation at a time by using this observation to upgrade already existing estimates, i.e. it generates a set of coefficients of a regression model, A_N, not by solving the matrix equation

$$A_N = \left(X_N{}^T X_N\right)^{-1}\left(X_N{}^T Y_N\right)$$

but by using only the last rows of arrays X_N and Y_N, i.e. $x(N) = [x_1(N),\ldots,x_m(N)]$ and $y(N)$, to update the previously obtained coefficients A_{N-1}:

$$A_N = A_{N-1} + \Delta[A_{N-1}, \; X(N), \; y(N)]$$

where $\Delta[.]$ is a specially defined increment, dependent on the initial conditions, A_{N-1}, and the most recent data $x(N) = [x_1(N),\ldots,x_m(N)]$ and $y(N)$. The first advantage of this approach is very obvious: the approach eliminates the need for storing large data arrays X_N and Y_N. The second advantage is even more important: the RLSM provides a mechanism for tracking the properties of time-varying processes by on-going correction of parameter estimates using the most recent data. The third advantage will become obvious later: RLSM eliminates the need for highly demanding computational procedure-matrix inversion.

Let us consider the mathematics behind the RLSM.

Recursive Least Squares Method Assume that several input variables, x_j, $j = 1,2,\ldots,m$, and one output variable, y, form the following relationship:

$$y(t) = \sum_{j=1}^{m} \alpha_j x_j(t)$$

where t is continuous time, and α_j, $j = 1, 2, .., m$ are unknown coefficients.

In order to estimate these coefficients, one should perform synchronous sampling of the input and output variables that would result in the following data arrays (it is expected that the sampling time step is constant)

$$X_N = \begin{bmatrix} x_1(1) & x_2(1) & x_3(1) & \cdots & x_m(1) \\ x_1(2) & x_2(2) & x_3(2) & \cdots & x_m(2) \\ \cdots & \cdots & \cdots & \cdots & \cdots \\ x_1(k) & x_2(k) & x_3(k) & \cdots & x_m(k) \\ \cdots & \cdots & \cdots & \cdots & \cdots \\ x_1(N) & x_2(N) & x_3(N) & \cdots & x_m(N) \end{bmatrix} \quad \text{and } Y_N = \begin{bmatrix} y(1) \\ y(2) \\ \cdots \\ y(k) \\ \cdots \\ y(N) \end{bmatrix}$$

where $k = 1,2,\ldots, N$ is the discrete-time index, and $N \gg m$ is the total number of measurements

It is known that the Least Square Method solution provides not the unknown coefficients, but rather their estimates

$$A_N = \begin{bmatrix} a_1 \\ a_2 \\ \cdots \\ a_j \\ \cdots \\ a_m \end{bmatrix} = \left(X_N^T X_N \right)^{-1} X_N^T Y_N$$

where ^{T} is the transpose symbol. Note that vector A has a subscript $_N$ that should remind the reader that estimates A_N are obtained using the total of N measurements of I/O variables.

It is also known that under some conditions estimates A_N have the following property:

$$\underset{N \to \infty}{\text{Lim}} \ A_N = \begin{bmatrix} \alpha_1 \\ \alpha_2 \\ \cdots \\ \alpha_j \\ \cdots \\ \alpha_m \end{bmatrix}$$

and therefore, it could be concluded that as number N increases, the "goodness" of the estimates also increases.

Assume the following situation: the total of N measurement have been accumulated, arrays X_N and Y_N have been formed, and estimates A_N have been computed. What to do when an additional measurement,

$$x(N+1) = \begin{bmatrix} x_1(N+1) \\ x_2(N+1) \\ \cdots \\ x_j(N+1) \\ \cdots \\ x_m(N+1) \end{bmatrix} \quad \text{and} \quad y(N+1)$$

has arrived? Should we upgrade arrays X_N and Y_N into

$$X_{N+1} = \begin{bmatrix} x_1(1) & x_2(1) & x_3(1) & \ldots & x_m(1) \\ x_1(2) & x_2(2) & x_3(2) & \ldots & x_m(2) \\ \ldots & \ldots & \ldots & \ldots & \ldots \\ \ldots & \ldots & \ldots & \ldots & \ldots \\ x_1(N) & x_2(N) & x_3(N) & \ldots & x_m(N) \\ x_1(N+1) & x_2(N+1) & x_3(N+1) & \ldots & x_m(N+1) \end{bmatrix} \quad \text{and}$$

$$Y_{N+1} = \begin{bmatrix} y(1) \\ y(2) \\ \ldots \\ \ldots \\ y(N) \\ y(N+1) \end{bmatrix}$$

and compute a new vector of estimates $A_{N+1} = \left(X_{N+1}{}^T X_{N+1}\right)^{-1} X_{N+1}{}^T Y_{N+1}$? Should we assume that a new set of I/O measurements will be received at time $N+2$, $N+3$, …. and continue this process at every time step? Obviously, the answer is "no" because of the following reasons:

1. We cannot designate an ever-growing memory for storing ever-growing arrays X_N and Y_N
2. We cannot perform the most time-consuming computation, matrix inversion, in real time, i.e. at every discrete time value $k = 1, 2, 3, \ldots, N, N+1, N+2, \ldots$
3. Statistics offers special recursive computational schemes for addressing the need for upgrading estimates utilizing continuously arriving "new" measurement data

Recall the recursive mean formula that demonstrates the power and numerical efficiency of recursive computations:

$$\overline{Z}_{N+1} = \overline{Z}_N + \frac{1}{N+1}\left(z(N+1) - \overline{Z}_N\right), \text{ where } N = 1, 2, 3, \ldots\ldots$$

It can be seen that the above formula is very attractive: it indicates that regardless of the number of measurements actually incorporated in the computation process, at any moment of discrete time, k, only the "old" estimate $\overline{Z}_{k-1}$, the "new" measurement z(k), and the discrete-time index "k" are to be stored. The computation results in the replacement of the "old" estimate $\overline{Z}_{k-1}$ by the "new" estimate $\overline{Z}_k$ that incorporates not only all previous measurements, z(1), z(2), …, z(k−1), but also z(k).

Now we will consider how to utilize this recursive approach for the Least Square Method computations. This has been accomplished by one of the brightest people in the world, Professor Kalman. First, recall the matrix inversion Lemma that Kalman formulated in 1960:

$$(\mathbf{A} + \mathbf{BCD})^{-1} = \mathbf{A}^{-1} - \mathbf{A}^{-1}\mathbf{B}(\mathbf{C}^{-1} + \mathbf{DA}^{-1}\mathbf{B})^{-1}\mathbf{DA}^{-1},$$

where $\mathbf{A}$, $\mathbf{C}$, and $\mathbf{C}^{-1} + \mathbf{DA}^{-1}\mathbf{B}$ are nonsingular square matrices. Now realize that if a single observation of the input variables is

$$x(k) = \begin{bmatrix} x_1(k) \\ x_2(k) \\ \cdots \\ x_j(k) \\ \cdots \\ x_m(k) \end{bmatrix}, \quad k = 1, 2, \ldots, N,$$

then array X_N could be defined as

$$X_N = \begin{bmatrix} x_1(1) & x_2(1) & x_3(1) & \cdots & x_m(1) \\ x_1(2) & x_2(2) & x_3(2) & \cdots & x_m(2) \\ \cdots & \cdots & \cdots & \cdots & \cdots \\ x_1(k) & x_2(k) & x_3(k) & \cdots & x_m(k) \\ \cdots & \cdots & \cdots & \cdots & \cdots \\ x_1(N) & x_2(N) & x_3(N) & \cdots & x_m(N) \end{bmatrix} = \begin{bmatrix} x(1)^T \\ x(2)^T \\ \cdots \\ x(k)^T \\ \cdots \\ x(N)^T \end{bmatrix}$$

Therefore $X_N{}^T X_N$ has the following expression using particular measurement vectors $x(k)$:

$$X_N{}^T X_N = \begin{bmatrix} x_1(1) & x_1(2) & x_1(3) & \cdots & x_1(N) \\ x_2(1) & x_2(2) & x_2(3) & \cdots & x_2(N) \\ \cdots & \cdots & \cdots & \cdots & \cdots \\ x_j(1) & x_j(2) & x_j(3) & \cdots & x_j(N) \\ \cdots & \cdots & \cdots & \cdots & \cdots \\ x_m(1) & x_m(2) & x_m(3) & \cdots & x_m(N) \end{bmatrix} \begin{bmatrix} x_1(1) & x_2(1) & x_3(1) & \cdots & x_m(1) \\ x_1(2) & x_2(2) & x_3(2) & \cdots & x_m(2) \\ \cdots & \cdots & \cdots & \cdots & \cdots \\ x_1(k) & x_2(k) & x_3(k) & \cdots & x_m(k) \\ \cdots & \cdots & \cdots & \cdots & \cdots \\ x_1(N) & x_2(N) & x_3(N) & \cdots & x_m(N) \end{bmatrix}$$

$$= \begin{bmatrix} \sum_{k=1}^{N} x_1(k)^2 & \sum_{k=1}^{N} x_1(k)x_2(k) & \cdots & \sum_{k=1}^{N} x_1(k)x_m(k) \\ \sum_{k=1}^{N} x_2(k)x_1(k) & \sum_{k=1}^{N} x_2(k)^2 & \cdots & \sum_{k=1}^{N} x_2(k)x_m(k) \\ \cdots & \cdots & \cdots & \cdots \\ \sum_{k=1}^{N} x_m(k)x_1(k) & \sum_{k=1}^{N} x_m(k)x_2(k) & \cdots & \sum_{k=1}^{N} x_m(k)^2 \end{bmatrix} = \sum_{k=1}^{N} x(k)x(k)^T$$

Similarly, $X_N{}^T Y_N$ could be expressed as $X_N{}^T Y_N = \sum_{k=1}^{N} x(k)y(k)$

Then rewrite the LSM equation as follows

$$A_N = \left(X_N^T X_N\right)^{-1} X_N^T Y_N = \left(\sum_{k=1}^{N} x(k)x(k)^T\right)^{-1} \sum_{k=1}^{N} x(k)y(k)$$

or

$$A_N = P_N \sum_{k=1}^{N} x(k)y(k) \text{ where } P_N = \left(X_N^T X_N\right)^{-1} = \left(\sum_{k=1}^{N} x(k)x(k)^T\right)^{-1}$$

Note that

$$P_N^{-1} = \sum_{k=1}^{N} x(k)x(k)^T = x(N)x(N)^T + \sum_{k=1}^{N-1} x(k)x(k)^T$$
$$= x(N)x(N)^T + P_{N-1}^{-1} \quad \neg$$

Also realize that

$$A_N = P_N \sum_{k=1}^{N} x(k)y(k) = P_N \left(x(N)y(N) + \sum_{k=1}^{N-1} x(k)y(k)\right) \quad \wedge$$

Since $A_N = P_N \sum_{k=1}^{N} x(k)y(k)$, it is obvious that $A_{N-1} = P_{N-1} \sum_{k=1}^{N-1} x(k)y(k)$ and therefore,

$$\sum_{k=1}^{N-1} x(k)y(k) = P_{N-1}^{-1} A_{N-1}$$

Since according to $\neg P_{N-1}^{-1} = P_N^{-1} - x(N)x(N)^T$, the above expression can be rewritten as

$$\sum_{k=1}^{N-1} x(k)y(k) = P_{N-1}^{-1} A_{N-1} = \left(P_N^{-1} - x(N)x(N)^T\right) A_{N-1}$$

Then $\wedge$ could be rewritten as

$$A_N = P_N \left(x(N)y(N) + \sum_{k=1}^{N-1} x(k)y(k)\right)$$
$$= P_N \left(x(N)y(N) + P_N^{-1} A_{N-1} - x(N)x(N)^T A_{N-1}\right)$$

or

$$A_N = P_N x(N) y(N) + P_N P_N^{-1} A_{N-1} - P_N x(N) x(N)^T A_{N-1} \quad \text{or}$$

$$A_N = A_{N-1} + P_N x(N) \left(y(N) - x(N)^T A_{N-1} \right) \quad \text{or}$$

$$A_N = A_{N-1} + K_N \left(y(N) - x(N)^T A_{N-1} \right) \quad \lor$$

where $K_N = P_N x(N)$

Expression $\lor$ constitutes a classical recursive formula. Indeed, it allows us to compute the "new" set of coefficients A_N on the basis of the "old" set A_{N-1} and the "new" measurement data, $x(N)$ and $y(N)$. However, it is too early to rejoice: note that matrix K_N is defined as

$$K_N = P_N x(N) = \left(\sum_{k=1}^{N} x(k) x(k)^T \right)^{-1} x(N)$$

i.e. formula $\lor$ does not eliminate the need for matrix inversion at every step of the procedure.

Let us utilize the Matrix Inversion Lemma:

$$(A + BCD)^{-1} = A^{-1} - A^{-1} B (C^{-1} + DA^{-1}B)^{-1} DA^{-1}$$

assuming that $A = P_{N-1}^{-1}$, $B = x(N)$, $C = I$, and $D = x^T(N)$:

$$P_N = \left(P_{N-1}^{-1} + x(N) x(N)^T \right)^{-1}$$

$$= P_{N-1} - P_{N-1} x(N) \left(I + x(N)^T P_{N-1} x(N) \right)^{-1} x(N)^T P_{N-1}$$

Use this result to define K_N:

$$K_N = P_N x(N) = P_{N-1} x(N) - P_{N-1} x(N) \left(I + x(N)^T P_{N-1} x(N) \right)^{-1} x(N)^T P_{N-1} x(N) \text{ or}$$

$$K_N = P_{N-1} x(N) \left\{ I - \left(I + x(N)^T P_{N-1} x(N) \right)^{-1} x(N)^T P_{N-1} x(N) \right\}$$

Let us treat $\left(I + x(N)^T P_{N-1} x(N) \right)^{-1}$ as a "common denominator" in the above expression, then

$$K_N = P_{N-1} x(N) \left\{ I + x(N)^T P_{N-1} x(N) - x(N)^T P_{N-1} x(N) \right\} \left(I + x(N)^T P_{N-1} x(N) \right)^{-1}$$

or

$$K_N = P_{N-1}x(N)\left(I + x(N)^T P_{N-1}x(N)\right)^{-1} \quad \Leftrightarrow$$

Note that expression $\Leftrightarrow$ does not contain matrix inversion: $I + x(N)^T P_{N-1}x(N)$ is a 1×1 matrix!

Now, using result $\Leftrightarrow$ let us rewrite

$$P_N = P_{N-1} - P_{N-1}x(N)\left(I + x(N)^T P_{N-1}x(N)\right)^{-1} x(N)^T P_{N-1}$$
$$= P_{N-1} - K_N x(N)^T P_{N-1} \text{ or } P_N = \left(I - K_N x(N)^T\right)P_{N-1}$$

Now the recursive parameter updating procedure (RLSM) can be defined as follows:

#1 Define Kalman gain: $K_N = P_{N-1}x(N)\left(I + x(N)^T P_{N-1}x(N)\right)^{-1}$

#2 Update parameters: $A_N = A_{N-1} + K_N\left(y(N) - x(N)^T A_{N-1}\right)$

#3 Provide matrix P for the next step: $P_N = \left(I - K_N x(N)^T\right)P_{N-1}$, $N = 1,2,3,\ldots$

and its block-diagram is given below in Fig. 2.2.

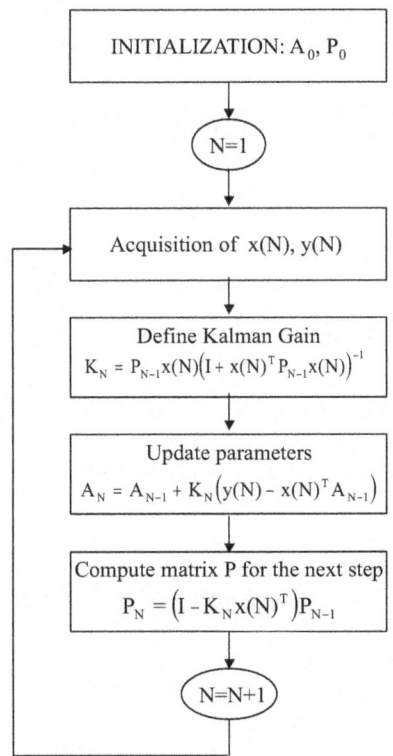

Fig. 2.2 RLSM procedure flowchart

It is well established that as the number of iterations approaches to infinity the parameter estimates A_N converge to the solution of the matrix equation of the LSM regardless of the initial conditions A_0 and P_0, i.e.

$$\lim_{N \to \infty} \left[A_{N-1} + K_N \left(y(N) - x(N)^T A_{N-1} \right) \right] = \lim_{N \to \infty} \left[(X_N^T X_N)^{-1} (X_N^T Y_N) \right]$$

In reality, especially when noise in the output is low, estimate A_N converges to the solution of the LSM equation very fast for any initial conditions. But how can these initial conditions be established?

Initial parameter estimates A_0 are the "best guess value" that are based on experience, intuition, preliminary experiments, or relevant equations of physics. Often, without any adverse consequences, it is assumed that $A_0 = [0, 0, \ldots, 0]^T$. The situation with the choice of matrix P_0 is more complex. It is often assumed that $P_0 = \alpha I$, where α is a positive constant and I is a unity matrix of the appropriate dimension. The choice of constant α has a profound effect on the convergence rate of the RLSM procedure—larger values of parameter α result in greater convergence rates of the RLSM estimation (see Example 2.8 below). Note that should the input/ output data be contaminated with noise, a much greater "overshoot" could be expected that potentially could crash the numerical process. Therefore, it is practical to test the procedure by choosing $\alpha = 0.01$, and if it would be found that a greater α value could be "afforded", the procedure should be tested with $\alpha = 0.1$, etc.

Estimation and Tracking It is known that for a large $N \gg 1$

$$A_N = A_{N-1} + K_N \left(y(N) - x(N)^T A_{N-1} \right) \approx (X_N^T X_N)^{-1} (X_N^T Y_N)$$

This property has a quite important implication: estimates A_N are based on entire arrays X_N and Y_N regardless of how large they are. On one hand, it seems to be an attractive property—utilization of large number of observations allows for "averaging out" the measurement noise. But there is an equally important consideration: most realistic processes exhibit parameter drift and their models, especially those developed for control applications, must reflect not the entire history of the process, but only its most current properties. The following graph in Fig. 2.3 provides an illustration of this concept.

First, note that $y^{MOD}(N,k)$ represents particular numerical values of the model output calculated as $y^{MOD}(N,k) = x(k)^T A_N$ where $A_N = (X_N^T X_N)^{-1}(X_N^T Y_N)$, $k = 1,2,3,\ldots, N$ is a discrete-time index, and N is the total number of observations utilized for the estimation of model parameters and consequently for the computation of the variance of the modeling error $\sigma_{ERR}^2(N)$. It could be realized the variance of the error increases as N decreases due to the inability to "average out noise", and increases as N increases due to the inability of the model to reflect current properties of the process that change during the observation period. Fortunately, $\sigma_{ERR}^2(N)$ has a minimum point, N^{OPT}, that clearly represents a rational

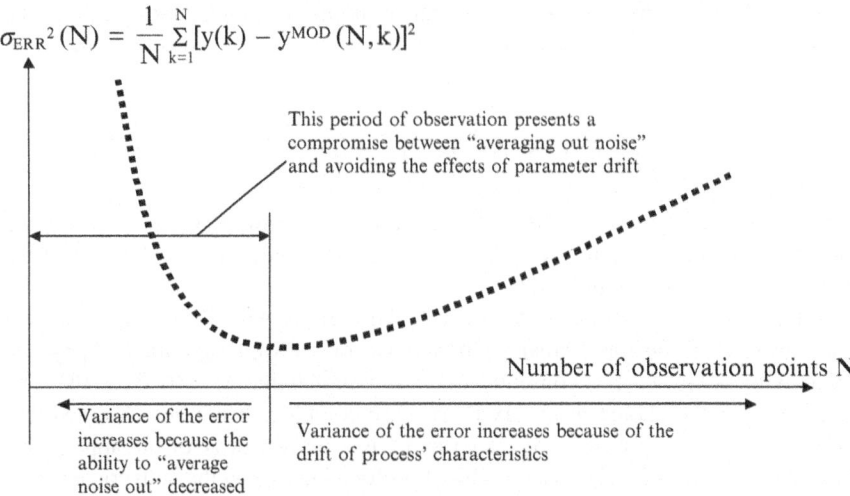

$$\sigma_{ERR}^2(N) = \frac{1}{N} \sum_{k=1}^{N} [y(k) - y^{MOD}(N,k)]^2$$

This period of observation presents a compromise between "averaging out noise" and avoiding the effects of parameter drift

Number of observation points N

Variance of the error increases because the ability to "average noise out" decreased

Variance of the error increases because of the drift of process' characteristics

Fig. 2.3 Variance vs number of observation points due to system parameter drift

compromise between two tendencies. Then a very important problem arises, how to assure that the model parameters estimated by the RLSM would be primarily based on the most recent N^{OPT} points and "forget" the impact of the preceding observations $x(k), y(k), k = 1,2,3,\ldots, N^{OPT}-1$?

This is achieved by the modification of the RLSM procedure by the introduction of the "forgetting factor" β, $0 < \beta \leq 1$:

#1 Define Kalman gain: $K_N = P_{N-1}x(N)\left(\beta + x(N)^T P_{N-1}x(N)\right)^{-1}$

#2 Update parameters: $A_N = A_{N-1} + K_N\left(y(N) - x(N)^T A_{N-1}\right)$

#3 Provide matrix P for the next step: $P_N = \frac{1}{\beta}\left(I - K_N x(N)^T\right)P_{N-1}$, $N = 1,2,3,\ldots$

It can be demonstrated that the above modification of the RLSM results in the estimates that are dominated by the last M observations, where M is known as the *memory* of the RLSM procedure, $M = \frac{1}{1-\beta}$. Choosing β such that $M = \frac{1}{1-\beta} = N^{OPT}$ is the way to respond to the above question. It converts the RLSM from being a conventional estimation tool ($\beta = 1$ and $M = \frac{1}{1-\beta} = \infty$) to a *tracking* tool, i.e. the procedure that is capable of tracking time-varying properties of the process on the basis of $M = N^{OPT}$ most recent process measurements.

Example 2.8 Investigation of the RLSM properties

The following CC code is utilized for generating the input, $x(k)$, and output, $y(k)$, data $k = 1,2,3,\ldots n$, (note that random noise is introduced in the output variable):

```
n=150;
p=(1,2,-3.6,8,-5);
p=diag(p);
x=randn(n,5)+ones(n,5)*p;
y=x*(5,-4,3,2.7,-6)';
for i=1:n & y(i,1)=y(i,1)+.1*randn(1,1) & end;
```

Then the RLSM with the forgetting factor is applied to estimate the "unknown" parameters of the model (note that "b" is the forgetting factor):

```
p=.01*iden(5);
b=1;
a=(0,0,0,0,0)';
for i=1:n & k=p*x(i,)'*(b+x(i,)*p*x(i,)')^-1 & a=a+k*(y(i,1)-x(i,)
*a)&;
aa(i,1)=a(1,1)' & p=(iden(5)-k*x(i,))*p/b & end;
a;
(x'*x)^-1*(x'*y);
plot(aa)
```

It can be seen that for comparison, the LSM-based estimates are being calculated, and the first component of the estimated parameters is plotted vs. the iteration index

$$A_{150} = \begin{bmatrix} 4.991 \\ -3.990 \\ 3.001 \\ 2.697 \\ -6.001 \end{bmatrix} \quad A^{LSM} = \begin{bmatrix} 4.994 \\ -3.993 \\ 3.003 \\ 2.697 \\ -6.004 \end{bmatrix} \quad A^{TRUE} = \begin{bmatrix} 5 \\ -4 \\ 3 \\ 2.7 \\ -6 \end{bmatrix}$$

For comparison, $A_{1000} = \begin{bmatrix} 5.000 \\ -4.001 \\ 3.002 \\ 2.703 \\ -5.997 \end{bmatrix}$

One can conclude that 150 iterations results in the estimates that are practically equivalent to the LSM estimates and are very close to the "true" parameter values.

The convergence of the estimate of the coefficient #1 is demonstrated below in Fig. 2.4.

Now let us investigate the effect of the forgetting factor on "averaging out noise" (using coefficient #1 of the model), as shown in Figs. 2.5 and 2.6.

Indeed, the increase of the forgetting factor leads to the increase of the memory thus reducing the effect of noise on the convergence of the estimates (Fig. 2.7).

Let us consider a very practical approach for the selection of the numerical value of the forgetting factor β. Assume that $x(k)$, $y(k)$, $k = 1,2,\ldots, N \gg 1$ represent input/output measurements of a process that exhibits parameter drift. It is also assumed

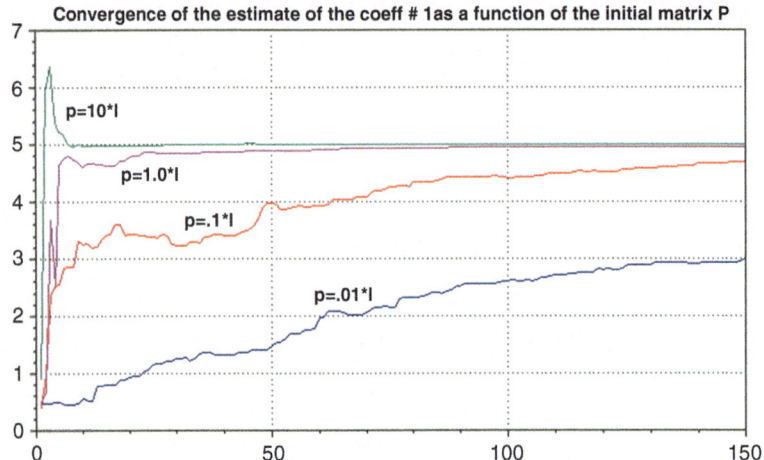

Fig. 2.4 Convergence of coefficient #1 as function of P

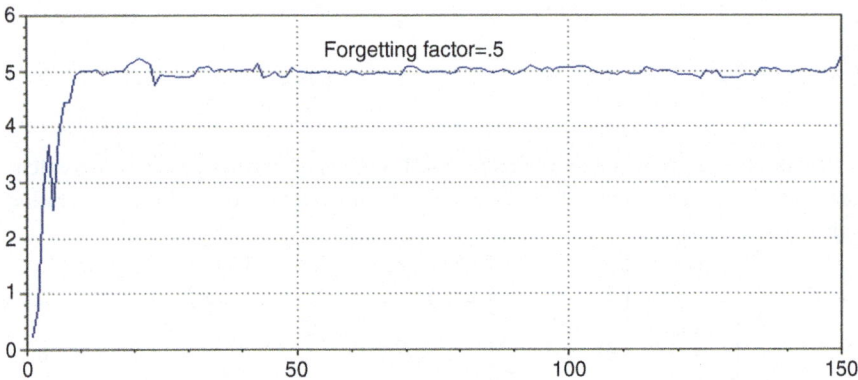

Fig. 2.5 Coefficient #1 convergence with forgetting factor 0.5

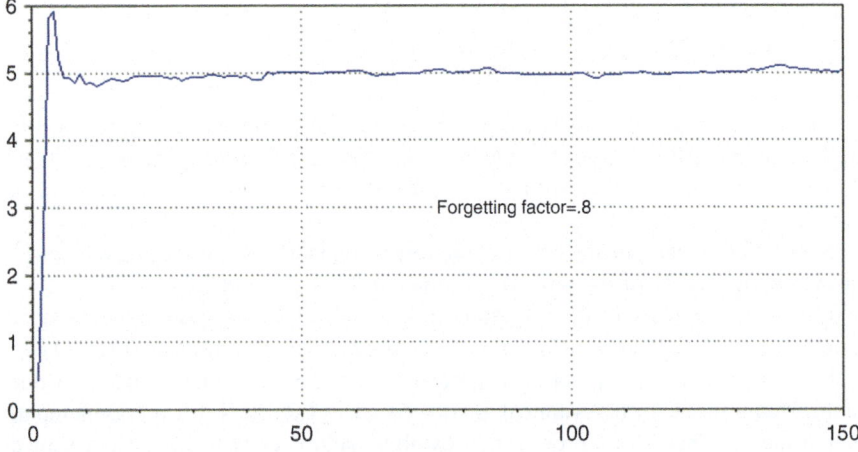

Fig. 2.6 Coefficient #1 convergence with forgetting factor 0.8

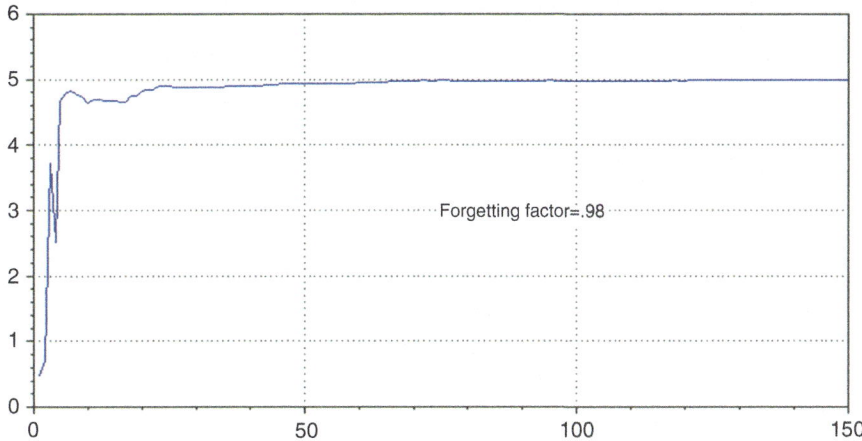

Fig. 2.7 Coefficient #1 convergence with forgetting factor 0.98

that measurements y(k), k = 1,2,... are contaminated with noise. It is proposed to select some value of the forgetting factor, $\beta = \beta_j$, run the RLSM procedure and simultaneously use the following recursive procedure for computing the coefficient of determination:

$$M_Y(k) = M_Y(k-1) + \frac{1}{k}[y(k) - M_Y(k-1)]$$

$$\sigma_Y^2(k) = \sigma_Y^2(k-1) + \frac{1}{k}\left([y(k) - M_Y(k)]^2 - \sigma_Y^2(k-1)\right)$$

$$\sigma_{MOD}^2(k) = \sigma_{MOD}^2(k-1) + \frac{1}{k}\left(\left[y(k) - x(k)^T A_k\right]^2 - \sigma_{MOD}^2(k-1)\right)$$

$$\eta(k) = \frac{\sigma_Y^2(k) - \sigma_{MOD}^2(k)}{\sigma_Y^2(k)}, \quad k = 1, 2, 3, \ldots\ldots$$

where A_k, k = 1,2,3,... are estimates of the model parameters generated by the RLSM with the forgetting factor $\beta = \beta_j$.

Plotting $\eta(k)$ as a function of the iteration index k = 1,2,3,... (see Example 2.9 below) reveals important insight on the selection of the forgetting factor:

1. In the situation when the forgetting factor is too small, the quality of the model represented by the coefficient of determination increases very fast, but the convergence of the RLSM is very much affected by noise because of the inability to "average noise out"
2. In the situation when the forgetting factor is too large, the quality of the model represented by the coefficient of determination improves slow and later even decreases because the RLSM cannot keep up with the parameter drift
3. In the situation when the forgetting factor is just right, the quality of the model, represented by the coefficient of determination, reaches and stays at its highest value

Properties of the RLSM Procedure

1. Under normal circumstances, the RLSM solution converges to the LSM, i.e. for $N \gg 1$

$$A_N = A_{N-1} + K(N)\big[y(N) - A_{N-1}{}^T X(N)\big] \approx \big(X_N{}^T X_N\big)^{-1} X_N{}^T Y_N$$

regardless of initial conditions A_0 and P_0

2. Noise in the input results in the biased parameter estimates
3. Noise in the output does not result in biased estimates. Estimation errors, caused by the output noise, could be controlled by increasing the memory size of the procedure, or by increasing the value of the forgetting factor.
4. The tracking ability of the RLSM procedure, i.e. the ability to generate accurate estimates of the time-varying parameters of the process, can be increased by reducing the memory size of the procedure, or by reducing the value of the forgetting factor.
5. Excessive cross correlation between input variables *prevents* RLSM from converging. This is the most common factor causing an RLSM procedure to fail.

Let us discuss the implications of property # 5. The Principal Component Analysis addresses numerical problems caused by cross correlation between input variables. It requires establishing the covariance matrix of the input vector, K_{XX}, defining its modal matrix, M, and reformulating the parameter estimation problem in terms of orthogonalized variables z:

$$B = \big(Z_N{}^T Z_N\big)^{-1} Z_N{}^T Y_N$$

where $Z_N = X_N M$. While Z_N is the array of N observations, particular observation $z(k)$ could be obtained from $x(k)$ as follows: $z(k) = M^T x(k)$, $k = 1,2,\ldots$ Moreover, according to the PCA, some of the components of the vector z could be eliminated, resulting in the input vector of lower dimension, $\hat{Z}$. For example component z_j, could be eliminated because of the appropriate eigenvalue λ_j of matrix K_{XX} being much smaller then other eigenvalues. Consequently the RLSM will be reduced to a very reliable (because of dealing with orthogonalized input variables) and fast (because of the reduced size of the problem) procedure:

$$K_N = P_{N-1}\hat{z}(N)\big(\beta + \hat{z}(N)^T P_{N-1}\hat{z}(N)\big)^{-1}$$

$$B_N = B_{N-1} + K_N\big(y(N) - \hat{z}(N)^T B_{N-1}\big)$$

$P_N = \dfrac{1}{\beta}\big(I - K_N\hat{z}(N)^T\big)P_{N-1}$, and $A_N = MB_N$, $N = 1, 2, 3, \ldots$.

Example 2.9 Selection of the optimal memory of the RLSM procedure. The following CC code presents the simulation study that performs this task:

```
n=1000;
p=(1,2,-3.6,8,-5);
p=diag(p);
x=randn(n,5)+ones(n,5)*p;
p=1*iden(5);
b=.999;
m=1/(1-b);
a=(0,0,0,0,0)';
my=0 & vy=0 & ve=0;
g=10
for i=1:n &;
c(1,1)=5+.2*g*i & c(2,1)=-4+g*.7*i &;
c(3,1)=3-g*.3*i & c(4,1)=2.7+g*.2*i &;
c(5,1)=-6+g*.13*i &;
y(i,1)=x(i,)*c+randn(1,1) &;
k=p*x(i,)'*(b+x(i,)*p*x(i,)')^-1 & a=a+k*(y(i,1)-x(i,)*a) &;
aa(i,1)=a(1,1)' & p=(iden(5)-k*x(i,))*p/b &;
err=y(i,1)-x(i,)*a;
my=my+(y(i,1)-my)/m & vy=vy+((y(i,1)-my)^2-vy)/m;
ve=ve+(err^2-ve)/m & nu(i,1)=(vy-ve)/vy & end;
plot(nu)
```

The results of this study, seen below in Fig. 2.8, indicate that the value of forgetting factor, $\beta = .9$, presents a compromise between the necessity to "average noise out" and perform successful tracking of time-variant process parameters.

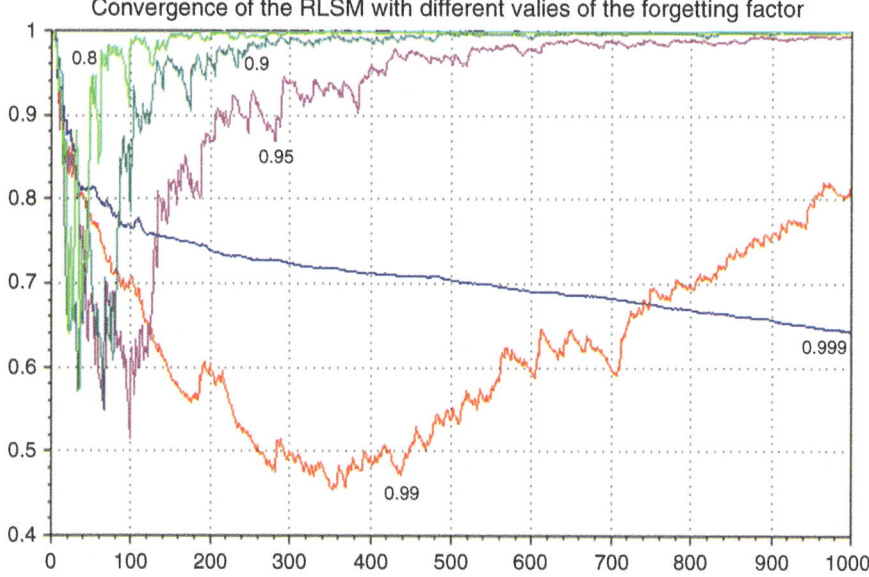

Fig. 2.8 RLSM convergence with different forgetting factors

Exercise 2.2

Problem 1 Utilize MATLAB to generate input and output variables as follows $(k = 1,2,\ldots,500)$:

$$x_1(k) = 5 + 3 \cdot \text{Sin}\,(17 \cdot k) + \text{Sin}\,(177 \cdot k) + .3 \cdot \text{Sin}\,(1771 \cdot k)$$
$$x_2(k) = 1 - 2 \cdot \text{Sin}\,(91 \cdot k) + \text{Sin}\,(191 \cdot k) + .2 \cdot \text{Sin}\,(999 \cdot k)$$
$$x_3(k) = 3 + \text{Sin}\,(27 \cdot k) + .5 \cdot \text{Sin}\,(477 \cdot k) + .1 \cdot \text{Sin}\,(6771 \cdot k)$$
$$y(k) = 2 \cdot x_1(k) + 3 \cdot x_2(k) - 0.4 \cdot x_3(k) + .5 \cdot x_1(k) \cdot x_3(k)$$
$$+x_2(k)^2 + 0.2 \cdot \text{Cos}(7671 \cdot k) - 0.1 \cdot \text{Sin}\,(17717 \cdot k)$$

Pretend that the "true" model configuration is unknown to you and assume that the model may include all linear and all second-order terms. Utilize LSM and the coefficient of determination technique to obtain the simplest but sufficiently accurate model. Document your work.

Problem 2 Implement RLSM with exponential forgetting in MATLAB. Test it by using $x_1(k)$, $x_2(k)$, and $x_3(k)$ as per Problem 1 and

$$y(k) = 8x_1(k) - 6x_2(k) + 5x_3(k) + 0.2 \cdot \text{Cos}(7671 \cdot k) - 0.1 \cdot \text{Sin}\,(17717 \cdot k),$$
$$k = 1, 2, \ldots 500$$

Assume $\beta = 1.$, $P(0) = .5*I$ and $A(0) = \begin{bmatrix} 0 \\ 0 \\ 0 \end{bmatrix}$. Plot the resultant parameter estimates.

Check if parameters A_N converge to the solution of the LSM problem. Compute the coefficient of determination of the "final" model $y^{MOD}(k) = x^T(k)A_{500}$

Problem 3 Redefine $y(k)$ to emulate a process with parameter drift

$$y(k) = [8 + .08k]x_1(k) - [6 - .04k]x_2(k) + [5 + .02k]x_3(k) + 0.2 \cdot \text{Cos}(7671 \cdot k)$$
$$- 0.1 \cdot \text{Sin}\,(17717 \cdot k),$$
$$k = 1, 2, \ldots 500$$

Assume $\beta = 1.$, $P(0) = .5*I$ and $A(0) = \begin{bmatrix} 0 \\ 0 \\ 0 \end{bmatrix}$. Plot the resultant parameter estimates

and the "true" time-dependent parameters. Compute the coefficient of determination of the "final" model $y^{MOD}(k) = x^T(k)A_{500}$, compare its value with the one obtained in Problem 2.

Problem 4 Redo Problem 3 with the forgetting factor $\beta < 1$. Plot the time-dependent "true" coefficients of the model and their RLSM estimates. Choose the optimal value of the forgetting factor that allows for the best tracking of the time-dependent "true" parameters.

2.4 Non-parametric Models. Cluster Analysis

Developing a Cluster Model Cluster analysis is a group of statistical techniques facilitating the detection of informative components of what could be a very extensive database. It is clear that this task cannot be accomplished without relevance to some decision-making or a classification problem. We will visualize the database as a combination of realizations of real status variables, X, and a binary class indicator, Q:

$$\{X(k), Q(k)\} = \{x_1(k), x_2(k), x_3(k), \ldots, x_n(k), Q(k)\} \tag{2.1}$$

where $k = 1, 2, 3, \ldots, N$ is the realization index, and $Q(k)$ can have only two alternative values, "**a**" or "**b**". Then the classification rule is established on the basis of some function defined in the X space, $F[X]$, such that, generally, $F[X] \leq 0$ for the majority of realizations when $Q(k) = {}''a''$ and $F[X] \succ 0$ for the majority of realizations when $Q(k) = {}''b''$, or in terms of conditional probabilities,

$$P\{F[k] \leq 0 \mid Q[k] = {}''a''\} \succ P\{F[k] \leq 0 \mid Q[k] = {}''b''\} \tag{2.2}$$

where $P\{A/B\}$ denotes the probability of event A subject to the occurrence of event B.

It is also understood that the classification problem does not have a unique solution, and there is a wide class of functions $F[X]$ that could satisfy condition (2.2) to a greater or a lesser extent. A simplification of the classification rule requires reducing the number of the components of vector X to the necessary minimum by choosing the smallest group of informative components that, in combination, allow for achieving reliable classification.

Selection of the informative components implies that contributions of particular groups of components of vector X to the classification are to be evaluated, and the most contributive group(s) be chosen for the definition of the classification rule. One can realize that in order to achieve the required discrimination power of the selection procedure, the groups must be small, and in order to consider combined effects of several variables must include at least two variables. Consider all possible combinations of two variables taken out of n, where n is the dimension of vector X. It could be said that the classification problem, originally defined in the space X, will now be considered on particular two-dimensional subspaces, $x_i \cap x_j$, where $i, j = 1, 2, \ldots, n$, and $i \neq j$.

Assume that the entire array of points, marked as "**a**" or "**b**", defined in the n-dimensional space X by the database (2.1), is projected on particular two-dimensional subspaces (planes). Let us visualize possible distributions of these projections. Figure A below in Fig. 2.9 illustrates a subspace that has no potential for the development of a classification rule due to the fact that points marked as "**a**" and "**b**" are distributed quite uniformly in this plane. The subspace of Figure B indicates a certain degree of separation between points "**a**" and "**b**" and,

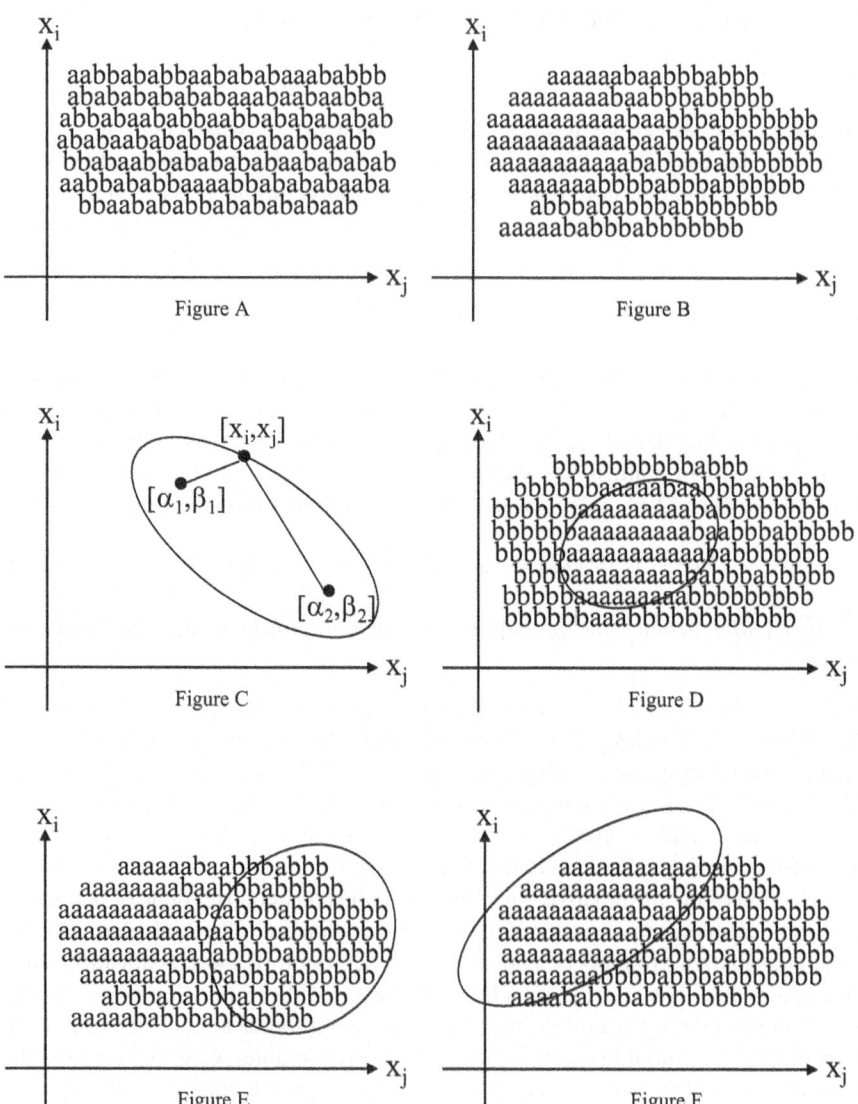

Fig. 2.9 Cluster analysis examples

therefore, should be viewed as informative. Figures D, E, F also illustrate possible cases of separation pattern in informative subspaces.

Consider the choice of some combination of the most informative two-dimensional subspaces $x_i \cap x_j$. This task could be performed by computing some informativity measure for every combination of two variables. The weighted average distance between points "**a**" and "**b**" in the particular subspaces,

$$\rho_{ij} = \frac{1}{N^a N^b \sigma_i \sigma_j} \sum_{k=1}^{N^a} \sum_{m=1}^{N^b} \sqrt{[x_i{}^a(k) - x_i{}^b(m)]^2 + [x_j{}^a(k) - x_j{}^b(m)]^2}$$

where σ_i are σ_j are standard deviations of variables x_i and x_j provides such a measure.

As shown in Figures D, E, F, a correlation ellipse, properly defined in the particular informative subspace, presents an ideal choice of the separating function. Figure C indicates that size, shape, position, and orientation of such an ellipse are defined by five parameters: coordinates of two focal points, $[\alpha_1, \beta_1]$, $[\alpha_2, \beta_2]$ and the constant δ, such that for any points of the ellipse, $[x_i, x_j]$, the following equation holds,

$$\sqrt{(x_i - \alpha_1)^2 + (x_j - \beta_1)^2} + \sqrt{(x_i - \alpha_2)^2 + (x_j - \beta_2)^2} = \delta \qquad (2.3)$$

Similarly, equations

$$\sqrt{(x_i - \alpha_1)^2 + (x_j - \beta_1)^2} + \sqrt{(x_i - \alpha_2)^2 + (x_j - \beta_2)^2} \leq \delta \qquad (2.3a)$$

$$\sqrt{(x_i - \alpha_1)^2 + (x_j - \beta_1)^2} + \sqrt{(x_i - \alpha_2)^2 + (x_j - \beta_2)^2} \succ \delta \qquad (2.3b)$$

represent any point $[x_i, x_j]$ within and outside the ellipse.

Consider the problem of the optimal definition of parameters $[\alpha_1, \beta_1, \alpha_2, \beta_2, \delta]$ of a correlation ellipse for a particular separation pattern in the plane comprising variables x_i and x_j. According to condition (2.2), this problem could be interpreted as the minimization of a loss function that includes a "penalty" for any point "a" outside the ellipse, $R^a(k) = R^a[x_i{}^a(k), x_j{}^a(k)]$, and a "penalty" for any point "b" within the ellipse, $R^b(k) = R^b[x_i{}^b(k), x_j{}^b(k)]$, i.e.

$$L(\alpha_1, \beta_1, \alpha_2, \beta_2, \delta) = \sum_{k=1}^{N^a} R^a(k) + \sum_{k=1}^{N^b} R^b(k) \qquad (2.4)$$

where N^a and N^b are number of points "a" and "b" in the database. Intuitively, these penalties are defined as follows:

$$R^a(k) = \begin{cases} 0, \text{ if point } [x_i{}^a(k), x_j{}^a(k)] \text{ satisfies condition (3a)} \\[2ex] \dfrac{1}{[x_i{}^a(k) - \mu_i{}^a]^2 + [x_j{}^a(k) - \mu_j{}^a]^2}, \text{ if point } [x_i{}^a(k), x_j{}^a(k)] \text{ satisfies condition (3b)} \end{cases}$$

and

$$R^b(k) = \begin{cases} 0, \text{ if point } \left[x_i^{\,b}(k), x_j^{\,b}(k)\right] \text{ satisfies condition (3b)} \\[2ex] \dfrac{1}{\left[x_i^{\,b}(k) - \mu_i^{\,b}\right]^2 + \left[x_j^{\,b}(k) - \mu_j^{\,b}\right]^2}, \text{ if point } \left[x_i^{\,b}(k), x_j^{\,b}(k)\right] \text{ satisfies condition (3a)} \end{cases}$$

where $[\mu_i^{\,a}, \mu_j^{\,a}]$ and $[\mu_i^{\,b}, \mu_j^{\,b}]$ are coordinates of the geometric centers of points "a" and points "b" distributed in the plain $x_i \cap x_j$. Such a choice of penalty functions places highest emphasis on the points in the immediate vicinity of geometric centers.

It could be seen that the loss function (2.4) is not only nonlinear but also discontinuous with respect to the unknown parameters of the separation ellipse $[\alpha_1, \beta_1, \alpha_2, \beta_2, \delta]$. Therefore our attempt to obtain the numerical values of these parameters by minimizing this loss function leads to a highly nonlinear multivariable optimization problem that does not have an analytical solution. Moreover, finding its global solution numerically would also be a very difficult task. Such an optimization problem presents an ideal application for a genetic optimization procedure that combines the advantages of both direct and random search. Application of genetic algorithms will be considered later in this course. It will result in the definition of an ellipse that indeed contains the largest possible number of points "a", N^{aa}, and the smallest possible number of points "b", N^{ab}. Then the "goodness" of the ellipse-based separating rule could be characterized by the following two quantities:

$$P_{in}\{a/a\} \approx \frac{N^{aa}}{N^a} \text{ and } P_{in}\{a/b\} \approx \frac{N^{ab}}{N^b} \tag{2.5}$$

representing the probabilities of a point "a" and a point "b" to be found *within* the ellipse, see Fig. 2.4.

Should we assume that the obtained classification rule, reflecting some compromise solution, could not be further improved? In our experience an alternative classification rule could be obtained by establishing an ellipse containing as many points "b", N^{bb}, and as few points "a", N^{ba}, as possible. This task is accomplished by the appropriate modification of the penalty functions. The resultant separating rule is characterized by:

$$P_{out}\{a/a\} \approx 1 - \frac{N^{ba}}{N^a} \quad \text{and} \quad P_{out}\{a/b\} \approx 1 - \frac{N^{bb}}{N^b} \tag{2.6}$$

representing the probabilities of a point "a" and a point "b" to be found *outside* the ellipse, see Fig. 2.5. The final selection of a separating rule implies the comparison of ratios

$$\rho_{in} = \frac{P_{in}\{a/a\}}{P_{in}\{a/b\}} \quad \text{and} \quad \rho_{out} = \frac{P_{out}\{a/a\}}{P_{out}\{a/b\}} \tag{2.7}$$

obtained for the "inside the ellipse" and "outside the ellipse" rules, and choosing the rule that results in the largest ρ value.

Finally, the proposed procedure implies

1. Definition of all two-dimensional subspaces of the space X
2. Computation of the informativity measure for every subspace $x_i \cap x_j$
3. Selection of a number M of the most informative subspaces
4. Selecting one of the M informative subspaces
5. Establishing the "inside the ellipse" classification rule by the application of a genetic optimization procedure, and computation of the ρ_{in} value for this subspace
6. Establishing the "outside the ellipse" classification rule by the application of a genetic optimization procedure, and computation of the ρ_{out} value for this subspace
7. Selection of the classification rule that has the largest ρ value for this subspace, and return to Step 4, until the list of informative subspaces will be exhausted.

Outcome Prediction Based on a Cluster Model Cluster analysis of the database results in the extraction and formalized representation of knowledge of various effects of the process outcome that constitutes a mathematical model of a special type. This model can be used for the prediction of the process outcome. The following mathematical framework is suggested for the attack detection scheme.

Assume that the preliminary cluster analysis utilizing the informativity criterion has resulted in the set of M two-dimensional informative subspaces. Then the set of M respective, either "inside the ellipse" or "outside the ellipse" classification rules, $R_i[X(k)]$, $i = 1,2,3,\ldots M$, has been developed. One can realize that each rule utilizes only those two components of vector X that constitute the ith informative subspace. For simplicity, assume that each classification rule is designed to return a negative value for the majority of points $X(k)$ marked by an "a". It is expected that every vector X^a, representing the outcome "a", and every vector X^b, representing the outcome "b", would satisfy only some classification rules but not all of them. Consider the following random events:

$$E_1 : R_1[X(k)] \leq 0 \cap R_2[X(k)] \leq 0 \cap R_3[X(k)] \leq 0 \cap R_4[X(k)] \leq 0 \cap \ldots R_M[X(k)] \leq 0$$

$$E_2 : R_1[X(k)] > 0 \cap R_2[X(k)] \leq 0 \cap R_3[X(k)] < 0 \cap R_4[X(k)] \leq 0 \cap \ldots R_M[X(k)] \leq 0$$

$$E_3 : R_1[X(k)] \leq 0 \cap R_2[X(k)] > 0 \cap R_3[X(k)] \leq 0 \cap R_4[X(k)] \leq 0 \cap \ldots R_M[X(k)] \leq 0$$

$$E_4 : R_1[X(k)] > 0 \cap R_2[X(k)] > 0 \cap R_3[X(k)] \leq 0 \cap R_4[X(k)] \leq 0 \cap \ldots R_M[X(k)] \leq 0$$

$$E_5 : R_1[X(k)] \leq 0 \cap R_2[X(k)] \leq 0 \cap R_3[X(k)] > 0 \cap R_4[X(k)] \leq 0 \cap \ldots R_M[X(k)] \leq 0$$

. . . .

$$E_L : R_1[X(k)] > 0 \cap R_2[X(k)] > 0 \cap R_3[X(k)] > 0 \cap R_4[X(k)] > 0 \cap \ldots R_M[X(k)] > 0 \tag{2.8}$$

representing specific combinations of classification rules satisfied by every vector $X(k)$. First, note that $L = 2^M$. Probabilities of events (2.8), evaluated separately for vectors X^a and X^b, constitute the following set of conditional probabilities instrumental for the outcome assessment procedure:

$$P\{E_i/a\} \text{ and } P\{E_i/b\}, \ i = 1, 2, \ldots, L \tag{2.9}$$

Now consider the utilization of the cluster model for the outcome assessment. Assume that the probability of the outcome "a" has some initial value, established according to the existing statistics, $\gamma[0]$, and therefore the probability of the outcome "b" is $\lambda[0] = 1 - \gamma[0]$.

Assume that vector $X(k) = [x_1(k), x_2(k), x_3(k), \ldots, x_n(k)]$ represents the set of the most process measurements. Numerical values of components of this vector, applied to the classification rules $R_i[X]$, $i = 1, 2, \ldots, M$, results in a particular combination of numerical values

$$R_1[X(k)], \ R_2[X(k)], \ R_3[X(k)], \ \ldots, \ R_M[X(k)]$$

that could be identified as the occurrence of one of the events (2.8), for example, E_j. Now the availability of conditional probabilities (2.9) facilitates the application of Bayesian approach for the re-evaluation of the probability of the outcome "a" (i.e. the probability of the point $X(k)$ to be marked by an "**a**") subject to the occurrence of the event E_j, $P\{a/E_j\}$. One can realize that unconditional probabilities, $P\{a\} + P\{b\} = 1$, therefore

$$P\{a/E_j\}P\{E_j\} = P\{E_j/a\}P\{a\} \text{ and } P\{E_j\} = P\{E_j/a\}P\{a\} + P\{E_j/b\}P\{b\},$$

and the required probability can be expressed as,

$$\begin{aligned} P\{a/E_j\} &= \frac{P\{E_j/a\}P\{a\}}{P\{E_j/a\}P\{a\} + P\{E_j/b\}P\{b\}} \\ &= \frac{\gamma[0] \cdot P\{E_j/a\}}{\gamma[0] \cdot P\{E_j/a\} + \lambda[0] \cdot P\{E_j/b\}} \end{aligned} \tag{2.10}$$

Computation (2.10) results in an updated value of the probability of attack, $P\{a/E_j\}$ that could be compared against some arbitrary defined threshold value. A message indicating that the expected outcome of the process is "a" could be issued if the probability of "a" exceeds the threshold. This completes one cycle of the proposed procedure. At the next cycle, the "prior" probabilities of the outcome "a" and the outcome "b" are defined as,

$$\gamma[k] = P\{a/E_j\} \text{ and } \lambda[k] = 1 - P\{a/E_j\}$$

and the new value of the vector,

$$X(k + 1) = [x_1(k + 1), x_2(k + 1), x_3(k + 1), \ldots, x_n(k + 1)]$$

is to be analyzed with the consequent computation of probabilities $\gamma[k + 1]$ and $\lambda[k + 1]$. This procedure, intended for continuous real-time application, is capable of providing timely and objective information to the operator providing mathematically justified support of his/her decisions.

Example 2.10 Application of the clustering approach. Consider a data array of 1000 measurements of the input and output variables of a complex industrial process. Each row of this array consists of the input vector $X(k) = [x_1(k)$, $x_2(k), \ldots x_{15}(k)]^T$ and the corresponding output variable $y(k)$, $k = 1, 2, \ldots 1000$ is the measurement index. The first 10 rows of the array could be seen below:

0.242726997 0.512632787 0.818820655 0.91894865 0.769801617 0.0557382442 0.792608202
0.365868866 0.158111617 0.381986767 0.0516359918 0.846051693 0.590528488
0.0123677235 0.864344776 **0.6**

0.0426672176 0.107910089 0.644807339 0.276805937 0.277515411 0.201748148 0.78121233
0.835532427 0.793120861 0.982161403 0.186749861 0.704956293 0.200024605 0.813628316
0.651033878 **0.3**

0.9261868 0.42181316 0.413958639 0.403003871 0.286023498 0.197082624 0.367629111
0.742382228 0.21803984 0.595475078 0.149892643 0.245591044 0.64862299 0.406341761
0.385907948 **0.3**

0.954634905 0.548775196 0.265029967 0.358599812 0.987211764 0.0684124753 0.80856663
0.57912004 0.270609707 0.137545109 0.720605493 0.216057882 0.284717888 0.25370416
0.00561538851 **0.6**

0.37790516 0.451865882 0.510103941 0.316611052 0.282071263 0.771665752 0.386009216
0.656956315 0.464612007 0.734265924 0.807381988 0.669486225 0.0551473089 0.860757768
0.755808294 **0.3**

0.870281875 0.827693462 0.0444668382 0.354088038 0.157880038 0.489894211 0.65190345
0.541297495 0.586609721 0.149126768 0.3736476 0.89600569 0.167734399 0.112052664
0.269221336 **0.3**

0.803052962 0.911425292 0.325179785 0.296805978 0.41806373 0.397285581 0.178858578
0.076258339 0.673950911 0.0937418342 0.518537939 0.0672319382 0.967123985
0.452468336 0.635296941 **0.3**

0.436127335 0.992104292 0.297061145 0.706475794 0.739071906 0.581460834 0.611842692
0.240244925 0.796370625 0.601304591 0.126385853 0.167113319 0.673507452 0.639618158
0.0626505017 **0.6**

0.967029989 0.873083532 0.915036321 0.0154176317 0.124073751 0.307632506 0.379356772
0.849460721 0.886274338 0.6125983 0.940086484 0.0336527638 0.602025151 0.236870512
0.0828597248 **0.3**

0.623347759 0.605348408 0.091186963 0.579296052 0.228726849 0.212254003 0.352962255
0.236755803 0.154763222 0.105398573 0.433777779 0.50333643 0.575454414 0.662479639
0.295345724 **0.3**

Note that the output variable $y(k)$ has only two distinct values, **.3** and **.6** that for all practical purposes could be interpreted as *A* and *B* correspondingly.

To discover unknown internal properties of the process leading to the outcome *A* or *B*, we will use cluster analysis implemented in computer program CLUSTER by the author (1995). This program extracts the prespecified number of the most informative subspaces and provides a very convincing visualization of the clustering phenomenon. The following is the printout of the program CLUSTER featuring six subspaces formed by input variables

$x_4 \& x_5$ - informativity criterion $= 5.243$, $x_4 \& x_9$ - informativity criterion $= 5.050$
$x_1 \& x_4$ - informativity criterion $= 5.045$, $x_2 \& x_4$ - informativity criterion $= 5.043$
$x_4 \& x_{10}$ - informativity criterion $= 5.041$, $x_4 \& x_8$ - informativity criterion $= 5.037$

```
INFORMATIVE SUBSPACE # 1:  X( 4) & X( 5) => CRITERION =      5.243

X( 4)   |0.1953E-02<----------------X( 5)---------------->0.9993E+00|
        |===========================================================|
0.0506>|   B  BB BBB   BB   B   B BBBX BBB   B   B   B B BABB B BBBB BBBBB|
0.1006>|B   BB BB  B   AA B B B A  B BB BBA    B BBBX B B BBBB B AXB |
0.1505>|  B      B B  B  BXBBXBXBBB BBBBBBBBB XB BBXBBBBXAB XA BBXX|
0.2005>|B B B  B BB  BBBB  B  B B BA  A BBB B      BB    A AB XA   B |
0.2504>|BB   B   BBB       B BBB B   BB BB AX A   B BBAA  XAX B |
0.3004>|BBB  B BB B    BBBBB B B BA  BBBB BBBXB  B   XA  B AAXAAXAX |
0.3503>|BB    BB BB    BB    BX B BBB X XA A AXBA  XX  AA A  AA A A|
0.4002>|B  BBB B B B BBBB BB B  BBX BXAAB A BAA  AB AABB     A  AAAA|
0.4502>|BB BB B  B  BB BBB  X BBB  BA A  A AAA AA  A A A    A A A A |
0.5001>|B B  BB  B    X  AXAXAB AB A   B AA  AA AA AAA A A AA A    |
0.5501>|    BBBBB A  BA  AAA    A A AABAA AA A AA A AAAAA  A AA AAA|
0.6000>|BBX B BB  A  BA AXB A  AA    A AAAAAA  A AAA A A    A  AAA|
0.6500>|   BB  AA   A AAA A A AAAA A A A    A A  AA    AAAAAAA AA|
0.6999>| B X B A  A A AAA AA AAAAA AA A AAAAAAA   A AAA AAAA   AA |
0.7499>| A X A   AA  A AA   AAA    AA  A A A AA AAA AAAAAAAA A A A |
0.7998>|AAAAA AA A  AAA    AA  AA AAAAA   A  AAA  A AAAAAAA A AAAA ||
0.8497>| A  A A A   AAA AAAAAAAA A AAA   AAA AAAA AAAAA  AA  AA A |
0.8997>|A  A AAAAAAA A  A AAA AA A A A    A AA  A AAA    A  AA  AA A|
0.9496>| A      A A  AAAA AA   AA  A AAAAAAAAA  A A A AAAA AA AAAAA|
0.9996>|AA  A A  AAA   AAAAA AA AA        A AAAAAAAA AAAA A     AA|
        |===========================================================|
        |0.1953E-02<----------------X( 5)---------------->0.9993E+00|


INFORMATIVE SUBSPASE # 2:  X( 4) & X( 9) => CRITERION =      5.050

X( 4)   |0.1325E-02<----------------X( 9)---------------->0.9999E+00|
        |===========================================================|
0.0506>|   BB  ABBBB  BBBBBB  BB BB XBB BB    B B  B B BBB BB  B  |
0.1006>|A   B BXBBB  BB B B     B   BBBBBB B A   B  XBBABBB XBBA  BBB |
0.1505>|BBB   X B BA BBBBB ABBB BB XABB  B  BB BBB ABXXABBBABB BBAXB|
0.2005>|  XB ABB XB   A      BB BBBABBBB  B BBBBB   AX   BX XB    |
0.2504>|B BBXB B BA B   B    B A B   BB B BB B B  B AX A    B A |
0.3004>|BBBB BXB BBB BBB X B  B BBBBA BB    ABBX AB  BXAA  XB  X AB|
0.3503>|B XBBB BBB  B  BXBA    XBB B A    A   B XA B   A   XA   XXA|
0.4002>|B  B AXBA    XXBB BBA  A B AB  B  BB   B  XXXB BB B  BABB A|
0.4502>|B  XBBX    BAB B B  XAA B  B   B B BAA AXA  X AA A  AB BAX |
0.5001>|   B   BAB X ABAAA  A  X  BA X A AAAAB    AA  X  A A B A XA|
0.5501>| ABB AA B  B XXA    B AAAA  AX ABX ABA A    AXA  AA  AA A  A|
0.6000>|AB A  BAAX AXX B A AX  AXAAXAA A AB XA A A  AAA AA AA A AA |
0.6500>| AA      AAXAA    X AA  A  A AAAAAAAAAA   AA A AA AAA A|
0.6999>|X AAAAAA AA  B X  A AAA  A AAAAAA  AA AA AAAA A  AAA   AA A A|
0.7499>| X  A AA AAAAA A  AA A AAAAA A  AA A   AA    A  AA AA   AA |
0.7998>| AAAA A   AA AA AAA   AAA AAA AA AA AA A AA A A AA AAAAAAAA |
0.8497>| A AAA A  AAAAA   AAA A AA  A A    AAAA AAAA   AAAA AA A AAA|
0.8997>|AA AA A  A A A   AAA A AAAA AAA     AA   A AAA A  AAAAA  |
0.9496>|  AA   AA A A  A   A  AAAAAAA  AA A A AAAAAAA AA     AAAA A|
0.9996>| A AAA   A  AA  AA  A A AAAA AA    A   AAAA AAA A A  AA A  A|
        |===========================================================|
        |0.1325E-02<----------------X( 9)---------------->0.9999E+00|
```

```
INFORMATIVE SUBSPASE # 3:  X( 1) & X( 4) => CRITERION =       5.045

X( 1)   |0.6811E-03<----------------X( 4)---------------->0.9996E+00|
        |================================================================|
0.0507>|BB    X BBBB  A   XXB  B   BABBAB   AB AA A XA A    AAAA      AA   |
0.1006>|  BBB   BBBAAB    BX BBBA  B  X     AXA AA  A AAA    AA  AA  A    |
0.1506>|   B BBXXBB     BBB     AABB B A AX  AAA AA AAA   AAA AA   A    A  |
0.2006>|BBB BB    BBB BBAB A    BXA    XA  AA   A A A  A A  AA AAAA  A   |
0.2505>| BBBBBBBB   BBB B BAXAAB  BXBB  XA  X      A AAAA  A A AAAAA|
0.3005>|BXBB BXBB   BB    AB BABBBB A A    AAABAA AAA AAA   AAA   A AAAA|
0.3504>|B  AXB BB       XB BB BB   X AB BAXBA  A AA AA   AAAA AAA A A  A|
0.4004>|B B X BBBBB ABX   AB X  BBB A    AAX AAA     A AA AAAA  AA   |
0.4504>| B X BAAAX BAA XX   BA BBA BA AAA   AAA BA   A X    A    AA AAAA|
0.5003>| B        A BAA  B       AA AX ABA  A  A AAA A AA   A  AA  |
0.5503>|BB  B    BBBBB    BXXB A A A  BA ABXX  AAA AA   AAA  A A A  AA|
0.6003>|BB B    BBBB    B  B B   BAAA XAAA   A  AAA AA  A A AA  AA A|
0.6502>|A    BB XAX   B   A B BBBAA B A AAAB AAAAAAA AAAAAAA A  AA AA|
0.7002>|    BBB  B A B BB A A BBX    AAAB AB  A A  AAAAAAAA A AA  AA|
0.7501>| BBB    B BB ABX BBA AX  AX  A    AA  AABAA A  AAAAA AAA AA|
0.8001>|BBBB BB B  BBB B BBBB   XBB B A  A B  AABAAA AAAAAAAAA      AA|
0.8501>|B BBB BXBXA B   BB   B A   B A BB    AA A AAA   A AA A AAA |
0.9000>|   B  AB B B AAB BAAB B  BA   ABBA    AAAA AAAA A A AA  A AA|
0.9500>|BBB B   X   B     BA    XXBAB  AA  AB AA A   AAAAAA  A A A  A|
0.9999>|BBBXBBXBX    B X BX A B AX    X  AAA AAAAA A A AA AA AAAAA   |
        |================================================================|
        |0.6811E-03<----------------X( 4)---------------->0.9996E+00|


INFORMATIVE SUBSPASE # 4:  X( 2) & X( 4) => CRITERION =       5.043

X( 2)   |0.6811E-03<----------------X( 4)---------------->0.9996E+00|
        |================================================================|
0.0501>|B B B  BBBXABB     B B  BAAABX XBBBA AA B AAAAA AA   AA AA A |
0.1001>| B B B AB A  BAXBA XAB X B B  A AA  A  A AAA  AAA A AAA    |
0.1501>|BB  B   BBB   B  BXABAAABB  BA   AAXA AAAA A AA AAA    A A A |
0.2000>| BBBB A X    BBXB BBA  A A   AA XAAA  XX A    AAAAAA A A AAAA|
0.2500>|BBBXBBBXX B  BBB AB X BAABA  AA BAA    ABA    AAAAAA AA AA A AA|
0.3000>|BB    BBBB BX  BBBBX AB A B   AB AAA A AAA  AAAA  AA A AAAA |
0.3499>|B  B   BX B B ABB B  ABBB   ABB A AA A   AAA A AAAAA AAA AA A|
0.3999>| B XBA BBXB XX BB  B   BA XABAAA  XAAAAA AA  AA AAA AAA  AA |
0.4499>| BB BBB BB B  X BXB A BAX XX   AA    AAA A AA A  AAAAAAAAAA A|
0.4998>| BB B  BBB  BB    XB   BA B   A  AAABAAX  AAA AA A A    A A |
0.5498>|   X   B AB    AAAX BB     XBA XA      A   A A A AAAAA|
0.5998>|BBB  B BAX AAB ABBAB XBB A    A   A    AAAA A AAA   AA AAAAAAA|
0.6497>|  BB  BBXB    XBBA BB BA     B AX  AA AA AAAA      AAAAAAAA|
0.6997>|B B B BX BB   A     XB B AAA AXXX AA AAA    AAAAAA  A    AA|
0.7497>|B   B  BB  B B  ABA AB XB B BA X AABA A AA AXAAAAAA A      A|
0.7996>|  BX  BXBBBB  AABBBB B X B B AAABA AA AA AA AA  AAAAAAA A AA|
0.8496>|B BBABBBBXB    B BB B B B X  A   A     AAA AAAA AAAAAA AAA A|
0.8996>|BB BXBAB B  BBAA B  B  AX A  A AABA A AAAA    AAAA AA  A |
0.9495>|XAB BB BBB B BBBABXBA   ABA      A A AA  AA AA AA  A  A   A |
0.9995>|BBBB BBB B  X   ABAB XB B  XXXAAA A A AA A  A AA  A AA   AAAA|
        |================================================================|
        |0.6811E-03<----------------X( 4)---------------->0.9996E+00|
```

```
INFORMATIVE SUBSPASE # 5:  X( 4) & X(10) => CRITERION =       5.041

X( 4)   |0.9187E-03<----------------X(10)---------------->0.9983E+00|
        |============================================================|
0.0506>|B BB     BBBBB  BB    BBBBBXBB   B BBBB  BBBBB  B B   BBX  B|
0.1006>|B    B B BBB BBBBBBBB AB XBBBBB BB        B  BXBBAB       XB B|
0.1505>|   BB B  B B B  BBBBBBAB B  BXBXB BXBAB AXABB BXBBBBBBABAB BA|
0.2005>|   BBBBB BB A  BBBBBBBBA A  BBABB  B      BX  BX X        BB|
0.2504>|   B B    X   A BB  BB  B B A  BX BABBAXA BB X     B    BBB |
0.3004>|    BBBABBBB BX ABXB   XABBX  ABXB X  B  B  X   B B XBAAXBBBBX|
0.3503>|  B   AABX B B  B XAABB XA  B  A XA X BA  A ABB B  X  BXBA BB|
0.4002>|XBXBA   ABBBBBX B  X   BABX  B A   XX  AB  B   B  BBBAAB    |
0.4502>|A XX BBB   A X       A  BABX   BBAA  X    AXA  B   AX B  BBA B|
0.5001>|AA XAB A B    AX BB  AB    A BA AX XB            BAAX   A A AA|
0.5501>|  A AA    AAAA AXBAAAA    AAAX  A A  BA  A B  A X BX  BAAAA  B|
0.6000>|    BXA AA  AA   A A  AAB BXAAA A  BB AAXB AAAA AAX   A BAAA|
0.6500>|A  AAA A AA A  A  AAAA A  AAAAAA XAAAAAAAXA  A AA    A  A A  |
0.6999>| A  A AA   A     A A A AA A   A  AAAA AAAAA AAA BA AABA  A AA|
0.7499>| AA A  AA  AAAA AA  AAAAA A AA    AA AAA A    AAAB   A AAAA|
0.7998>|     AA  AAA AAAAAAAAAAAAA   A AAAA AAA AAAAA AAA AA A AA  A|
0.8497>|AA     AAAAAA AAAAAAAAAA  AA   AA AAAAAAAAAA A  AA A  AAA A A|
0.8997>|AA  A AA A  A AA AAA      A  A AA AAA  AAAAAA AAAA     A  A |
0.9496>| AA  AAAAA A AA AA A AAAAAA  A     A  AA   A  A  AAAAA AAAA|
0.9996>|  AA    A  AA A AAA     A  AA A A A A  AA AAA AAAAAA AA A  A|
        |============================================================|
        |0.9187E-03<----------------X(10)---------------->0.9983E+00|


INFORMATIVE SUBSPASE # 6:  X( 4) & X( 8) => CRITERION =       5.037

X( 4)   |0.1869E-02<----------------X( 8)---------------->0.9999E+00|
        |============================================================|
0.0506>|BBBB    BBB BB  BA     BB BBBBBB    XBB BBB B   BBBB  BBBBB B|
0.1006>|XBBABB   XB  BBB B    B   B BAB   BB B   BABBB X B BBB B BBA|
0.1505>|BBX BBBB   B B ABBBBAB B BABB  B BXXBAXX B  BB B X AB BB A  B|
0.2005>|B  B  B  XAB B    B  BB    X  X B B BXB  B BBBXBA AB B   B  B |
0.2504>|   X  A XBB B B  B  B B B A     B    BB B X  BX A A      BBB|
0.3004>|A   BBBB    B BXBAX AX XB X   B    AXA B  XA  BB BX BBBBX BB|
0.3503>|BA X  BBXBA  AA A B XX A  B A X    A   X   AX  B  XBBXBAX  |
0.4002>|  B    BBB  B AB X  A XXB  A ABBBB  BXABBAB B B B AAXA A BBB|
0.4502>|ABB  B   AAB AABX X   AA A B X  BAX     BBBB A XB B     B B|
0.5001>|A   B  A  A AAAABAB   AX B AA   A B A A  B  XX AAA A  A B B|
0.5501>| XA   XA  BA A A  AXB    AB A XAA AXA A  B    A  A AA A AAABA|
0.6000>|AA  A XA AAXA XA BA A AAAAB  ABABA A  X  X  A B   X XAAA AAA|
0.6500>|    AA AA A A AAA AA A AA A AAAA A AAAAA  A  AXXAAAAAAA  A  A|
0.6999>|AA AAAA AAA AA  AA A     AAAA BAAAB  AA A AXAAAAAA A  A|
0.7499>|    AA  AAAAA  AAA A A A  AAA A   A AA    AA XA A A    AAA A|
0.7998>|AAA AAAA A A  A AAAA AAAA AA AA AAA AAA AA   AA   AAA A  A A|
0.8497>|AAAA AAAA   A   A A  A AA  AAA AAAA A AAAAAAA AAA   AAAAAAAAAA|
0.8997>|A AA A AA A AA A A AAAA A AA A A AAA AAA    AAAAAAA A  AAA|
0.9496>|A AA AAA AAA AAAA   AAAAA   AA AA  AA   A AAAA  AAA A A AA|
0.9996>|  AA  AA    A A  AA  A  AAA AAA AAA   A AAA AAAAA   AA A |
        |============================================================|
        |0.1869E-02<----------------X( 8)---------------->0.9999E+00|
```

The results of cluster analysis provide important insight into the understanding of the nature of the process, and most important, enable the process operator to establish the recommended "operational window" in terms of the process variables. In addition, as shown below it facilitates the predicting of the process outcome.

Example 2.11 Prediction of the process outcome based on the cluster analysis. The clustering pattern featured in Example 2.10 indicates that separating rules could be defined simply by straight lines rather than ellipses. A straight line, $x_1 + a_1x_2 + a_0 = 0$, is to be defined in each subspace in such a way that the majority of points A and B be located on different sides of this line, i.e. $x_1 + a_1x_2 + a_0 > 0$ for the majority of coordinates of points A and $x_1 + a_1x_2 + a_0 < 0$ for the majority of coordinates of points B. The development of such lines for the chosen subspaces presents a simple manual task.

For simplicity, consider the first three subspaces of the previous example formed by process variables x_4&x_5, x_4&x_9 and x_1&x_4. First consider the subspace x_4&x_5. As shown in the figure below, the separating line $x_4 + a_1x_5 + a_0 = 0$ can be drawn by inspection:

```
X( 4)    |0.1953E-02<-----------------X( 5)----------------->0.9993E+00|
         |=================================================================
0.0506>|    B   BB BBB    BB    B   B BBBX BBB   B    B   B B BABB B BBBB BBBBB|
0.1006>|B    BB BB    B    AA B B B A   B BB BBA     B BBBX B B BBBB B AXB
0.1505>|   B        B  B   B   BXBBXBXBBB BBBBBBBBB XB BBXBBBBXAB  XA BRX  ②
0.2005>|B B B    B BB   BBBB    B   B B BA   A BBB B        BB      A AB XA    B|
0.2504>|BB       B      BBB          B BBB B    BB BB AX A   B BBAA   XAX B |
0.3004>|BBB   B BB B      BBBBB B B BA  BBBB BBBXB  B    XA   B AAXAAXAX |
0.3503>|BB     BB BB      BB     BX B BBB X XA A AXBA   XX   AA A  AA A A|
0.4002>|B   BBB B B B BBBB BB B   BBX BXAAB A BAA   AB AABB     A  AAAA|
0.4502>|BB BB B   B   BB BBB   X BBB  BA A  A AAA AA   A A A    A   A A A |
0.5001>|B B   BB   B    X  AXAXAB AB A   B AA  AA AA AAA A A AA A  |
0.5501>|     BBBBB A  BA   AAA    A A AABAA AA A AA A AAAAA   A AA AAA|
0.6000>|BBX B BB   A  BA AXB A  AA   A AAAAAA   A AAA A A    A  AAA|
0.6500>|     BB  AA   A AAA A A AAAA A A A     A A  AA     AAAAAAA AA|
0.6999>①  B X B A  A A AAA AA AAAAA AA A AAAAAAA   A AAA AAAA    AA |
0.7499>| A X A    AA  A AA     AAA     AA  A A A AA AAA AAAAAAAA A A A |
0.7998>|AAAAA AA A  AAA    AA  AA AAAAA    A  AAA  A AAAAAAA A AAAA |
0.8497>|  A  A A  A    AAA AAAAAAAA A AAA    AAA AAAA AAAAA  AA  AA A |
0.8997>|A  A AAAAAAA A  A AAA AA A A A    A AA  A AAA    A  AA  AA A|
0.9496>|  A      A A AAAA AA    AA  A AAAAAAAAA  A A A AAAA AA AAAAA|
0.9996>|AA  A A  AAA    AAAAA AA AA        A AAAAAAAA AAAA A      AA|
         |=================================================================
         |0.1953E-02<-----------------X( 5)----------------->0.9993E+00|
```

Point 1 coordinates: x_4=.7, x_5=0   **Point 2 coordinates: x_4=.15, x_5=1.0**

This results in the computation of coefficients a_1 and a_0 based on equations:

$$x_4 + a_1x_5 + a_0 = 0 \quad \text{or}$$
$$.7 + a_1 \cdot 0 + a_0 = 0 \quad \rightarrow a_0 = -.7$$
$$.15 + a_1 - .7 = 0 \quad \rightarrow a_1 = .55$$

Now perform the same task for subspace x_4&x_9, see the figure below. The equations are:

$$x_4 + a_1x_9 + a_0 = 0 \quad \text{or}$$
$$.6 + a_1 \cdot 0 + a_0 = 0 \quad \rightarrow a_0 = -.6$$
$$.3 + a_1 - .6 = 0 \quad \rightarrow a_1 = .3$$

```
X( 4)   |0.1325E-02<------------------X( 9)---------------->0.9999E+00|
        |=================================================================|
0.0506>|     BB    ABBBB   BBBBBB  BB BB XBB BB     B B  B B BBB BB  B   |
0.1006>|A   B BXBBB  BB B B    B   BBBBBB B A   B  XBBABBB XBBA  BBB|
0.1505>|BBB    X B BA BBBBB ABBB BB XABB  B   BB BBB ABXXABBBABB BBAXB|
0.2005>|  XB ABB XB   A      BB BBBABBBB  B BBBBB   AX   BX XB       |
0.2504>|B B BBXB B BA B    B     B  A B   BB B BB B B  B AX  A    B A  |
0.3004>|BBBB BXB BBB BBB X B  B BBBBA BB     ABBX AB  BXAA   XB   X AB |
0.3503>|B XBBB BBB   B  BXBA    XBB B A     A    B XA B   A    XA   XXA|
0.4002>| B  B AXBA    XXBB BBA  A B AB  B  BB   B  XXXB BB B  BABB A|
0.4502>|B  XBBX    BAB B B  XAA B  B   B B BAA AXA  X AA A  AB BAX   |
0.5001>|   B   BAB X ABAAA  A  X  BA X A AAAAB    AA  X  A A B A XA|
0.5501>| ABB AA B  B XXA    B AAAA  AX ABX ABA A   AXA  AA  AA A  A|
0.6000>|AB A  BAAX AXX B A AX  AXAAXAA A AB XA A A  AAA AA AA A AA   |
0.6500>| AA    AAXAA     X AA  A A  AAAAAAAAAA      AA A AA AAA A|
0.6999>|X AAAAAA AA  B X  A AAA  A AAAAAA  AA AA AAAA A  AAA  AA A A|
0.7499>| X  A AA AAAAA A  AA A AAAAA A AA A   AA    A  AA AA   AA  |
0.7998>| AAAA A   AA AA AAA  AAA AAA AA AA AA A AA A A AA AAAAAAAA   |
0.8497>| A AAA A  AAAAA   AAA A AA  A A   AAAA AAAA   AAAA AA A AAA|
0.8997>|AA AA A  A A  A   AAA A AAAA AAA     AA   A AAA A  AAAAA  |
0.9496>|    AA    AA A A  A   A  AAAAAAA  AA A A AAAAAAA AA    AAAA A|
0.9996>| A AAA   A  AA  AA  A A AAAA AA   A    AAAA AAA A A  AA A  A|
        |=================================================================|
        |0.1325E-02<------------------X( 9)---------------->0.9999E+00|
```

Point 1 coordinates: $x_4=.6$, $x_9=0$ **Point 2 coordinates: $x_4=.3$, $x_9=1.0$**

Finally, the analysis of the third chosen subspace x_1&x_4 is as follows:

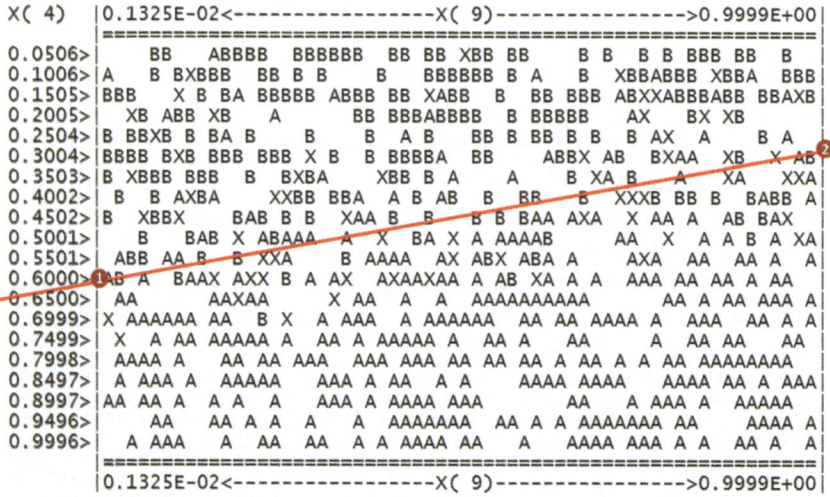

```
X( 1)   |0.6811E-03<------------------X( 4)---------------->0.9996E+00|
        |=================================================================|
0.0507>|BB   X BBBB A  XXB B  BABBAB  AB AA A XA A   AAAA    AA       |
0.1006>|  BBB  BBBAAB   BX BBBA B  X    AXA AA  A AAA   AA  AA  A  |
0.1506>|  B BBXXBB    BBB   AABB B A AX  AAA AA AAA  AAA AA  A    A  |
0.2006>|BBB BB   BBB BBAB A   BXA   XA  AA   A A A  A A  AA AAAA  A  |
0.2505>| BBBBBBBB   BBB B BAXAAB  BXBB  XA X       A AAAA  A A AAAAA|
0.3005>|BXBB BXBB  BB   AB BABBBB A A  AAABAA AAA AAA  AAA   A AAAA|
0.3504>|B  AXB BB     XB BB BB  X AB BAXBA  A AA AA  AAAA AAA A A  A|
0.4004>|B B X BBBBB ABX  AB X  BBB A    AAX AAA    A AA AAAA  AA   |
0.4504>| B X BAAAX BAA XX  BA BBA BA AAA  AAA BA  A X    A   AA AAAA|
0.5003>| B        A BAA B     AA AX ABA  A  A AAA A AA  A  AA   |
0.5503>|BB  B    BBBBB   BXXB A A A  BA ABXX  AAA AA   AAA  A A A  AA|
0.6003>|BB B     BBBB    B  B B  BAAA XAAA  A  AAA AA  A A AA  AA A|
0.6502>|A    BB XAX   B  A B BBBAA B A AAAB AAAAAAA AAAAAAA A  AA AA|
0.7002>|    BBB  B A B BB A A BBX    AAAB AB  A A  AAAAAAAA A AA  AA|
0.7501>| BBB   B BB ABX BB A AX  AX A   AA  AABAA A  AAAAA AAA AA  |
0.8001>|BBBB BB B  BBB B BBBB   XBB B A  A B  AABAAA AAAAAAAAA      AA|
0.8501>|B BBB BXBXA B   BB  B A   B A BB     AA A AAA   A AA A AAA |
0.9000>|    B  AB B B A B BAAB B  BA   ABBA   AAAA AAAA A A AA  A AA|
0.9500>|BBB B   X  B   BA    XXBAB  AA  AB AA A   AAAAAA  A A A  A|
0.9999>|BBBXBBXBX   B X BX A B AX   X  AAA AAAAA A A AA AA AAAAA    |
        |=================================================================|
        |0.6811E-03<------------------X( 4)---------------->0.9996E+00|
```

Point 1 coordinates: $x_1=1$, $x_4=.2$ **Point 2 coordinates: $x_1=.05$, $x_4=.6$**

$$x_1 + a_1x_4 + a_0 = 0 \quad \text{or}$$
$$1.0 + a_1 \cdot .2 + a_0 = 0$$
$$.05 + a_1 \cdot .6 + a_0 = 0 \quad \rightarrow a_1 = 2.375$$
$$1.0 + 2.375 \cdot .2 + a_0 = 0 \quad \rightarrow a_0 = -1.475$$

Now, the existence of the separating conditions, enables us to subject the original data array $X(k) = [x_1(k), x_2(k),...x_{15}(k)]^T$, $y(k)$, $k = 1,2,...1000$, to probabilistic analysis by using program ANALYST written by V. Skormin (2015). This analysis implies

– The calculation of probabilities of occurrence of two possible outcomes of the process $P(A)$ and $P(B)$,

– The detection of the occurrence of events $E_1, E_2, E_3, E_4, E_5, E_6, E_7, E_8$ for each measurement of process variables by computing the following functions for each $k = 1,2,...,1000$

$$\varphi_1(.) = x_4(.) + a_1x_5(.) + a_0$$
$$\varphi_2(.) = x_4(.) + a_1x_9(.) + a_0$$
$$\varphi_3(.) = x_1(.) + a_1x_4(.) + a_0$$

then E_1 is defined as $\varphi_1 < 0$ & $\varphi_2 < 0$ & $\varphi_3 < 0$
E_2 is defined as $\varphi_1 < 0$ & $\varphi_2 < 0$ & $\varphi_3 \geq 0$
E_3 is defined as $\varphi_1 < 0$ & $\varphi_2 \geq 0$ & $\varphi_3 < 0$
E_4 is defined as $\varphi_1 < 0$ & $\varphi_2 \geq 0$ & $\varphi_3 \geq 0$
E_5 is defined as $\varphi_1 \geq 0$ & $\varphi_2 < 0$ & $\varphi_3 < 0$
E_6 is defined as $\varphi_1 \geq 0$ & $\varphi_2 < 0$ & $\varphi_3 \geq 0$
E_7 is defined as $\varphi_1 \geq 0$ & $\varphi_2 \geq 0$ & $\varphi_3 < 0$
E_8 is defined as $\varphi_1 \geq 0$ & $\varphi_2 \geq 0$ & $\varphi_3 \geq 0$

– The calculation of conditional probabilities of events $E_1, E_2, E_3, E_4, E_5, E_6$, E_7, E_8 subject to outcome A and outcome B, i.e.

$P(E_1/A), P(E_2/A), P(E_3/A), P(E_4/A), P(E_5/A), P(E_6/A), P(E_7/A), P(E_8/A)$,
$P(E_1/B), P(E_2/B), P(E_3/B), P(E_4/B), P(E_5/B), P(E_6/B), P(E_7/B), P(E_8/B)$

The following is the printout of the program ANALYST:

```
P[A]=0.625  P[B]=0.375

P[E1/A]=0.014 P[E2/A]=0.000 P[E3/A]=0.018 P[E4/A]=0.018
P[E5/A]=0.056 P[E6/A]=0.038 P[E7/A]=0.035 P[E8/A]=0.821


P[E1/B]=0.711 P[E2/B]=0.091 P[E3/B]=0.045 P[E4/A]=0.078
P[E5/B]=0.043 P[E6/B]=0.016 P[E7/B]=0.000 P[E8/B]=0.016
```

These results offer a probabilistic basis for the prediction of the process outcome based on the immediate measurements of the input variables. The importance of this problem is justified by possible considerable delays in receiving the information on process outcome due to the process dynamics and delays in measurement channels (the end product may be transported to a laboratory for testing). Note that according to available data, initial probabilities of the process outcome are $P(A) = .625$ and $P(B) = .375$, i.e. $P(A) + P(B) = 1$.

Assume that the table below contains two consequent measurements of the process variables:

x_1	x_2	x_3	x_4	x_5	x_6	x_7	x_8	x_9	x_{10}	x_{11}	x_{12}	x_{13}	x_{14}	x_{15}
.470	.827	.044	.354	.158	.490	.652	.541	.887	.149	.373	.896	.168	.112	.269
.703	.911	.325	.296	.418	.397	.179	.076	.674	.094	.518	.067	.967	.452	.635

Compute functions representing the separating lines for the first and second measurements:

$$\varphi_1(.) = x_4(.) + a_1 x_5(.) + a_0 = .354 + .158 \cdot .55 - .7 = -.259 < 0$$
$$\varphi_2(.) = x_4(.) + a_1 x_9(.) + a_0 = .354 + .887 \cdot .3 - .6 = -.02 > 0$$
$$\varphi_3(.) = x_1(.) + a_1 x_4(.) + a_0 = .470 + .354 \cdot 2.375 - 1.475 = -.164 < 0$$

This result is indicative of event E_3. According to Bayes' formula,

$$P(A/E_3) = \frac{P(E_3/A) \cdot P(A)}{P(E_3/A) \cdot P(A) + P(E_3/B) \cdot P(B)} = \frac{.018 \cdot .625}{.018 \cdot .625 + .045 \cdot .375} = .402$$

Consequently, $P(B/E_3) = 1 - .402 = .598$

Let us upgrade this result further based on the second measurement:

$$\varphi_1(.) = x_4(.) + a_1 x_5(.) + a_0 = .296 + .418 \cdot .55 - .7 = -.174 < 0$$
$$\varphi_2(.) = x_4(.) + a_1 x_9(.) + a_0 = .295 + .674 \cdot .3 - .6 = -.102 < 0$$
$$\varphi_3(.) = x_1(.) + a_1 x_4(.) + a_0 = .703 + .296 \cdot 2.375 - 1.475 = -.069 < 0$$

This result indicates event E_1. According to Bayes' formula assuming $P(A) = .402$ and $P(B) = .598$

$$P(A/E_1) = \frac{P(E_1/A) \cdot P(A)}{P(E_1/A) \cdot P(A) + P(E_1/B) \cdot P(B)} = \frac{.014 \cdot .402}{.014 \cdot .402 + .711 \cdot .598} = .013$$

Consequently, $P(B/E_1) = 1 - .013 = .987$

The analysis indicates that almost certainly the outcome of the process is expected to be A, i.e. output variable $y = .6$

2.5 Non-parametric Models. Singular-Value Decomposition as a Tool for Cluster Analysis

Singular Value Decomposition (SVD) is a standard numerical tool readily available to a modern engineer. It could be utilized for various process control applications providing a dependable feature extraction, cluster analysis and pattern matching techniques.

It is understood that monitoring of any complex phenomenon (situation, process, structure, data set, etc.) results in a set of real numbers $\{x_1, x_2, \ldots, x_j, \ldots, x_N\}$ that could be "folded" in an $m \times n$ matrix,

$$
A = \begin{bmatrix}
x_1 & x_2 & \ldots & x_n \\
x_{n+1} & x_{n+2} & \ldots & x_{2n} \\
\ldots & \ldots & \ldots & \ldots \\
x_{N-1} & x_N & \ldots & 0
\end{bmatrix}
$$

In many cases the matrix is obtained by the very nature of the data monitoring system, recall array X_N from the LSM material.

It is known that SVD allows for the representation of such a matrix in the form

$$
A = \sum_{j=1}^{K} \sigma_j P_j Q_j^T
$$

where

$K = m$, if $m > n$, and $K = n$ otherwise,

$\sigma_j, j = 1, 2, \ldots, K$ are non-negative singular values of matrix A,

$P_j, j = 1, 2, \ldots, K$ are m-dimensional left singular vectors of matrix A, and

$Q_j, j = 1, 2, \ldots, K$ are n-dimensional right singular vectors of matrix A

Since most SVD procedures list singular vectors in order of their decreasing contribution to the sum, matrix A could be approximated by a partial sum of first $L < K$ terms of the above expression and the accuracy of such an approximation increases as L approaches K. It could be said that if matrix A represents a particular process, then its left and right singular vectors carry essential features of this process and could be utilized for the purpose of process identification.

Consider the task of determining the degree of similarity between a process represented by matrix A and a process represented by matrix B of the same dimension. Traditionally, this task requires comparing these matrices on the element-by-element basis thus for high-dimensional matrices results in a computationally intensive task. This task, however, could be carried out by measuring matrix B not against the entire matrix A, but against features extracted from matrix A, i.e. by computing and analyzing scalars

$$w_j = P_j^T B Q_j, j = 1, 2, \ldots, K$$

It is important that one can reduce this task to a very short list of features thus utilizing only the most important features of matrix A

$$w_j = P^T_j B Q_j, j = 1, 2, \ldots, L < K$$

In most practical situations it is sufficient to have $L = 2$, 3, 4. For the purpose of visualization, introduce two-dimensional feature space, where coordinates (features) are the first two scalars,

$$w_1 = P^T_1 B Q_1 \text{ and } w_2 = P^T_2 B Q_2$$

Then the location of point $\{w_1, w_2\}$ in the feature space would represent the degree of commonality between matrices A and B, or between the two complex phenomena (situations, processes, structures, data sets, etc.) that they represent. It is important to realize that the largest absolute numerical values of quantities w_i, $i = 1, 2, \ldots$ are expected when $A = B$.

Figure 2.10 depicts the degree of commonality between process A and processes B, C, and D represented by points labeled as ^{AB}, ^{AC} and ^{AD} in the feature space. It clearly indicates that the degree of commonality between processes A and B is different from the one between A and C and A and D.

Figure 2.11 represents the results of matching various processes, B, C, D, E, F, G, H, L to process A using two pairs of its singular vectors to compute quantities w_1 & w_2.

It is obvious that according to the resultant clustering pattern there are three groups of processes of similar nature: Group 1: A and B, Group 2: C, E, F, and Group 3: D, G, H, L.

Note that two-dimensional feature space is considered only to facilitate the visualization. In order to perform computer-based analyses, the dimensionality of the feature space could be increased, or more than one purposely chosen subspaces

Fig. 2.10 Commonality plot for A to B, C, and D

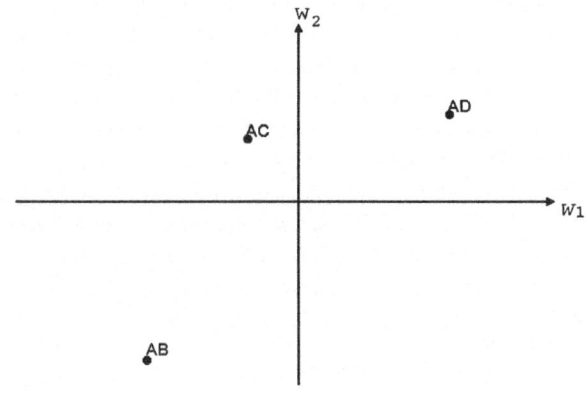

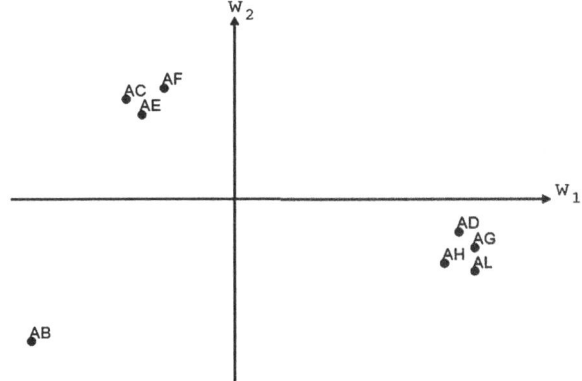

Fig. 2.11 Commonality plot for A to B, C, D, E, F, G, H, and L

of the feature space could be utilized. Then the analysis would address specific properties of the processes and enhance the robustness of the recognition procedure.

Originally, feature extraction and cluster analysis were developed for computer-based pattern recognition and, historically, are known as machine learning. The SVD approach is highly instrumental for both supervised and unsupervised leaning.

In the case of ***supervised learning***, assume that $m \times n$-dimensional matrices, matrix A, matrix B, and matrix C are known to be representatives of three different classes of processes. The application of SVD procedure to respective matrices will result in the extraction of the following left and right singular vectors of these matrices,

$$P_{Aj}, \; Q_{Aj}, \; P_{Bj}, \; Q_{Bj}, \; P_{Cj}, \; Q_{Cj}, \; j = 1, 2, \ldots,$$

where notations are self-explanatory. Now assume that an $m \times n$-dimensional matrix S representing unknown process could be subjected to the following computation

$$\delta^A = \sqrt{\left(P_{A1}^{\;T} S Q_{A1}\right)^2 + \left(P_{A2}^{\;T} S Q_{A2}\right)^2 + \ldots,}$$

where $P_{Ai}^{\;T} S Q_{Ai}$, represents the ith component, $i = 1,2,\ldots$ of the feature space where matrix S in question is compared with known matrix A, and ^{T} is the transpose symbol. In the similar fashion matrix S is compared with known matrix B and known matrix C

$$\delta^B = \sqrt{\left(P_{B1}^{\;T} S Q_{B1}\right)^2 + \left(P_{B2}^{\;T} S Q_{B2}\right)^2 + \ldots}$$

$$\delta^C = \sqrt{\left(P_{C1}^{\;T} S Q_{C1}\right)^2 + \left(P_{C2}^{\;T} S Q_{C2}\right)^2 + \ldots}$$

Then process (matrix) S can be easily recognized as a process of class A if $\delta^A > \delta^B$ and $\delta^A > \delta^C$.

The **unsupervised learning** could be implemented as follows. Assume that a complex process has a number of distinctive states, A, B, C, ... Assume that a sequence of measurement vectors $X(1)$, $X(2)$, ..., $X(k)$, ..., $X(N)$, where $X(k) = [x_1(k) \quad x_2(k) \quad .... \quad x_m(k)]^T$ is taken over a sufficiently long period of time representing all distinctive states of process. This data could be compiled in a $m \times n$ matrix X_N defined as

$$X_N = \begin{bmatrix} x_1(1) & x_2(1) & \cdots & x_m(1) \\ x_1(2) & x_2(2) & \cdots & x_m(2) \\ \cdots & \cdots & \cdots & \cdots \\ x_1(N) & x_2(N) & \cdots & x_m(N) \end{bmatrix} = \begin{bmatrix} X(1)^T \\ X(2)^T \\ \cdots \\ X(N)^T \end{bmatrix}$$

Subject matrix X_N to SVD that results in singular values, left and right singular vectors, σ_j, P_j, Q_j, $j = 1, 2, ...$ Then utilizing the first two right singular vectors Q_1 and Q_2 of matrix X_N compute the two coordinates of the feature space for each $k = 1, 2, ..., N$,

$$w_1(k) = X(k)^T \frac{1}{\sigma_1} Q_1, \quad \text{and } w_2(k) = X(k)^T \frac{1}{\sigma_2} Q_2$$

These coordinates are highly instrumental for making a distinction between the process states A, B, C, ... represented by vectors $X(k)$, $k = 1, 2, ..., N$. Indeed, in the feature space points with coordinates $[w_1(k), w_2(k)]$, $k = 1, 2, ..., N$ tend to form clusters consistent with process states (classes) A, B, C, ... This reality enables one to discover the very existence of specific classes $A, B, C, ...$ and define them in the factor space utilizing the clustering phenomenon. It should be noted that the feasibility and dependability of the described procedure is dependent on the characteristics of noise in the information channels forming vectors $X(k)$, $k = 1, 2, ..., N$.

Consequently, one can conclude that the analytical technique, described herein, could facilitate solution of a wide class of feature extraction, and feature-based recognition tasks.

Illustration of Unsupervised Learning at Different Signal-to-Noise Ratio A simulation study described below demonstrates a successful application of the SVD to unsupervised learning. The following numerical procedure generates matrices, representing four different processes, A, B, C, and D. Each matrix is contaminated by noise, note that two signal-to-noise ratios were considered: .5 and 1.0. The results are highly robust with respect to signal-to-noise ratio and are displayed in Figs. 2.12 and 2.13. The application of the unsupervised learning procedure results in a clear clustering phenomenon that could be used for the process recognition/identification.

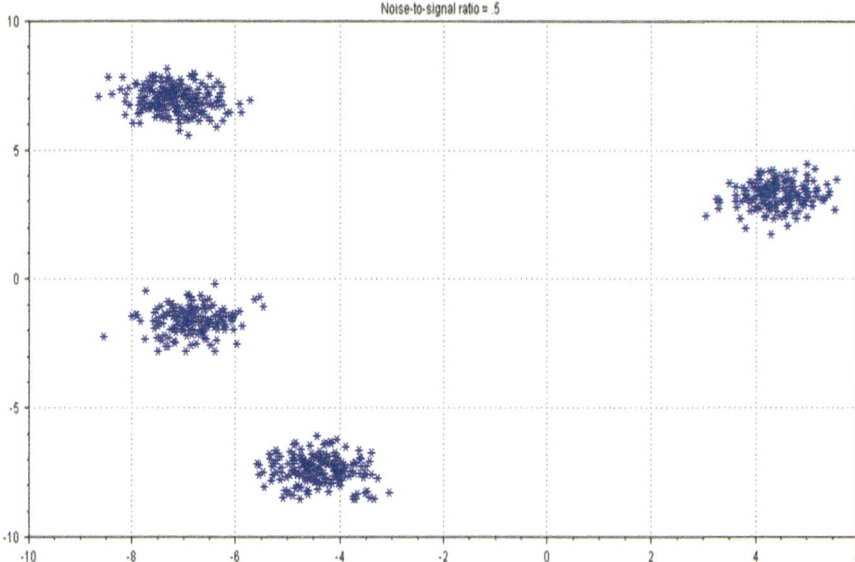

Fig. 2.12 SVD for noise to signal ratio 0.5

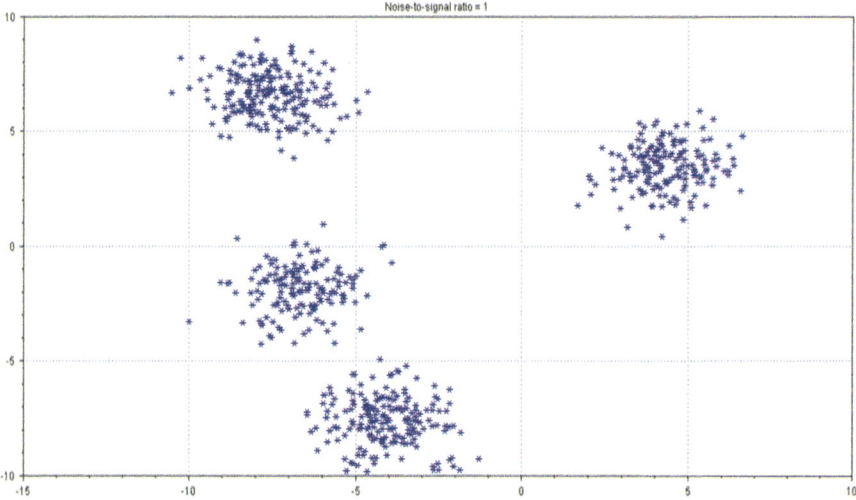

Fig. 2.13 SVD for noise to signal ratio 1.0

```
N=350
m=100
f=.5
a=randn(m,1)
b=randn(m,1)
```

```
c=randn(m,1)
d=randn(m,1)
for k=1:N
x=randn(1,1)
if x>=0
mat=a+randn(m,1)*f
else
mat=b+randn(m,1)*f
end
h(,k)=mat
end
for k=1:N
x=randn(1,1)
if x>=0
mat=c+randn(m,1)*f
else
mat=d+randn(m,1)*f
end
h(,k+N)=mat
end
hh=h'
(u,S,v)=svd(hh)
svl1=u(,1)
svl2=u(,2)
svr1=v(,1)
svr2=v(,2)
mm=2*N
for k=1:mm
x=hh(k,)
w1(k,1)=x*svr1
w2(k,1)=x*svr2
end
plot(w1,w2,'*')
```

One can enhance the unsupervised learning procedure by utilizing more than two components of the feature space, perhaps, several subspaces of the feature space. A special recognition procedure utilizing the classification rules established by unsupervised learning could be developed.

Exercise 2.3 The purpose of this assignment is to add the conventional cluster analysis and analysis based on singular value decomposition to students' professional toolbox. Both techniques will significantly extend students' ability to develop useful mathematical models of complex industrial processes.

Problem 1 Cluster analysis for modeling and outcome prediction of a multivariable process with real inputs and discrete-event output.

Task 1. Run program GENERATOR.EXE that will create a customized array of 1000 measurements of a complex industrial process with 15 input variables and a discrete event-type output. The value of the output variable could be only .3 (outcome A) or .6 (outcome B). This data will be recorded in file TSMD.DAT

Task 2. Run program CLUSTER.EXE that will process data in the file TSMD. DAT by

a) assessing the impact of every combination of two process variables on the process outcome (i.e. their contribution to the classification problem)

b) creating file SUBSPACE.DAT that will contain the prespecified number (three is recommended) of the most informative subspaces and depict the clustering phenomena in each subspace

Task 3. At each subspace $x_1 \& x_2$ establish the separating line, $x_1 + k \cdot x_2 + q = 0$ such that the majority of points A and majority of points B be located on different sides of the separating line. This task requires one to evaluate coefficients k and q for each of the three subspaces.

Task 4. Run program ANALYST.EXE that will

a) request pairs of variable indexes for each of the three most informative subspaces

b) request values of coefficients k and q for every separating line

c) detect events E_1–E_8 that may occur at each row of TSMD.DAT defined as follows:

E_1: $\varphi_1 < 0$ & $\varphi_2 < 0$ & $\varphi_3 < 0$
E_2: $\varphi_1 < 0$ & $\varphi_2 < 0$ & $\varphi_3 \geq 0$
E_3: $\varphi_1 < 0$ & $\varphi_2 \geq 0$ & $\varphi_3 < 0$
E_4: $\varphi_1 < 0$ & $\varphi_2 \geq 0$ & $\varphi_3 \geq 0$
E_5: $\varphi_1 \geq 0$ & $\varphi_2 < 0$ & $\varphi_3 < 0$
E_6: $\varphi_1 \geq 0$ & $\varphi_2 < 0$ & $\varphi_3 \geq 0$
E_7: $\varphi_1 \geq 0$ & $\varphi_2 \geq 0$ & $\varphi_3 < 0$
E_8: $\varphi_1 \geq 0$ & $\varphi_2 \geq 0$ & $\varphi_3 \geq 0$

where

$\varphi_j = x_i^j + k^j \cdot x_k^j + q^j$ is the equation of the separating line of the j – th subspace, comprizing two process variables, $x_i^j \& x_k^j$

d) display self explanatory results of the probabilistic analysis of the TSMD.DAT data

Task 5. Given two successive measurements of the input variables of the process:

[0.633 0.768 0.502 0.814 0.371 0.057 0.618 0.141 0.279 0.363 0.538 0.662 0.844 0.322 0.298]^T

[0.255 0.018 0.318 0.967 0.884 0.640 0.813 0.488 0.402 0.067 0.235 0.257 0.413 0.393 0.849]^T

Utilize Bayes' approach to define probabilities of the process outcomes based on these measurements.

Problem 2 Familiarizing with properties of SVD and its applications for cluster analysis.

Task 1. Find the singular value decomposition (SVD) facility in the MATLAB library. Learn how to access first and second, left and right singular vectors, L_1, R_1 and L_2, R_2

Task 2. Find the random matrix generator (RMG) facility in the MATLAB library. Learn how to access the resultant matrix.

Task 3. Use RMG to generate a 400×600 matrix A. Use SVD to obtain singular vectors, L_{A1}, R_{A1} and L_{A2}, R_{A2} of this matrix.

Task 4. Use MATLAB code to generate ten "noisy versions" of matrix A, i.e. $\underline{A}(i) = A + .3 \cdot \Delta(i)$, where each $\Delta(i)$ is a unique random 400×600 matrix generated by the RMG. For each matrix $\underline{A}(i)$ compute $w_{A1}(i) = L_{A1}^T \cdot \underline{A}(i) \cdot R_{A1}$ and $w_{A2}(i) = L_{A2}^T \cdot \underline{A}(i) \cdot R_{A2}$, $i = 1,2,\ldots,\overline{10}$

Task 5. Use RMG to generate a 400×600 matrix B. Use MATLAB code to generate ten "noisy versions" of matrix B, i.e. $\underline{B}(i) = B + .3 \cdot \Delta(i)$, where each $\Delta(i)$ is a unique random 400×600 matrix generated by the RMG. For each matrix $\underline{B}$ (i) compute $w_{B1}(i) = L_{A1}^T \cdot \underline{B}(i) \cdot R_{A1}$ and $w_{B2}(i) = L_{A2}^T \cdot \underline{B}(i) \cdot R_{A2}$, $i = 1,2,\ldots,\overline{10}$. (Note: use the same singular vectors L_{A1}, R_{A1} and L_{A2}, R_{A2} as in Task 4)

Task 6. Use RMG to generate a 400×600 matrix C. Use MATLAB code to generate ten "noisy versions" of matrix C, i.e. $\underline{C}(i) = C + .3 \cdot \Delta(i)$, where each $\Delta(i)$ is a unique random 400×600 matrix generated by the RMG. For each matrix $\underline{C}(i)$ compute $w_{C1}(i) = L_{A1}^T \cdot \underline{C}(i) \cdot R_{A1}$ and $w_{C2}(i) = L_{A2}^T \cdot \underline{C}(i) \cdot R_{A2}$, $i = \overline{1},2,\ldots,10$. (Note: use the same singular vectors L_{A1}, R_{A1} and L_{A2}, R_{A2} as in Task 4)

Task 7. Use MATLAB plotting facility to place points with coordinates [$w_{A1}(i)$, $w_{A2}(i)$], $i = 1,\ldots10$, marked with "■", and points with coordinates [$w_{B1}(i)$, $w_{B2}(i)$], $i = 1,\ldots10$, marked with "□", and points with coordinates [$w_{C1}(i)$, $w_{C2}(i)$], $i = 1,\ldots10$, marked with "●" on the $w_1 w_2$ plane. Print out the resultant figure. Comment on the capabilities of SVD.

Solutions

Exercise 2.1: Problem 1

For each subset of the X and Y matrices, the coefficients of A were calculated using the following LSM procedure:

$$A = (X^T \times X)^{-1} \times X^T \times Y$$

The following is the set of coefficient obtained from respective number of data points and their "true" values:

30	100	200	500	True
1.9811	1.9996	1.9967	1.9998	2.0000
3.0406	2.9106	3.0463	2.9823	3.0000
−2.0476	−2.0071	−2.0084	−2.0020	−2.0000
4.9867	5.0290	4.9838	5.0058	5.0000

It could be seen that the greater the number of data points, the more accurate the approximation of A coefficients is.

Exercise 2.1: Problem 2

For each coefficient of the suggested model, a 95% confidence interval was built based on the error of the model and the respective diagonal elements of the covariance matrix, i.e. q_{ii}

$$Q = K_{xx}^{-1} = \left[\frac{1}{N} \times X^T \times X \right]^{-1}$$

The half-width for each confidence interval was calculated as

$$\Delta_i = t(\alpha = .025, N = 300) \times \sigma_E^2 \times \sqrt{\frac{q_{ii}}{N}}$$

The 95 % confidence interval for the model coefficient a_1 is 1.9561 to 2.0435 and the "true" a_1 is 2, so the true parameter lies within the interval.

The 95 % confidence interval for the model coefficient a_2 is 2.4562 to 3.5084 and the "true" a_2 is 3, so the true parameter lies within the interval.

The 95 % confidence interval for the model coefficient a_3 is −2.0745 to −1.9296 and the "true" a_3 is −2, so the true parameter lies within the interval.

The 95 % confidence interval for the model coefficient a_4 is 4.832 to 5.1797 and the "true" a_4 is 5, so the true parameter lies within the interval.

Exercise 2.1: Problem 3

Given the following set of input values:

$$\tilde{X} = [2.5 \quad 3 \quad -6.3 \quad 10]$$

and matrix $Q = K_{xx}^{-1} = \left[\frac{1}{N} \times X^T \times X \right]^{-1}$

The 95 % confidence interval half-width was calculated as

$$\Delta = t(\alpha = .025, N = 300) \times \sigma_E^2 \times \sqrt{\frac{\tilde{X}^T \times Q \times \tilde{X}}{N}}$$

The 95 % confidence interval for output Y is 75.9734 to 77.2616 and the "true" Y is 76.31, so the true Y lies within the interval.

Exercise 2.1: Problem 4

The required covariance matrices are:

$K_{xx} =$

3.0840	0.8460	1.1580	−0.4340	0.3430
0.8460	10.9000	4.0120	0.1040	1.4420
1.1580	4.0120	6.2690	0.0010	1.5430
−0.4340	0.1040	0.0010	3.4250	0.2660
0.3430	1.4420	1.5430	0.2660	0.6770

$K_{xy} =$
12.9700
−19.7400
7.2130
−8.6490
0.6100

$K_{noise} =$

0.7500	0	0	0	0
0	1.6600	0	0	0
0	0	0.9600	0	0
0	0	0	0.2600	0
0	0	0	0	0.1100

The model parameters were estimated by the following procedure:

$$A = (K_{xx})^{-1} \times K_{xy}$$

The calculated model parameters A are:
3.0896
−2.4294
1.5077
−2.0313
1.6105

The estimation errors were calculated with the following procedure:

$$Error_{noise} = \left[(K_{xx} - K_{noise})^T \times (K_{xx} - K_{noise})\right]^{-1} \times K_{xy} - (K_{xx}^T \times K_{xx})^{-1} \times K_{xy}$$

The parameter estimation errors cause by this known noise are:

0.8977
−0.4328
0.0208
0.0433
0.4748

Exercise 2.1: Problem 5

First, matrix Z was calculated from matrix X and matrix W.

$$Z = X \times W$$

"Artificial" coefficients B were calculated from Z.

$$B = (Z^T \times Z)^{-1} \times Z^T \times Y$$

Then, the variance of Y was calculated and the variance for each B was calculated.

$$\sigma_{z(i)}^2 = \lambda_i - (\overline{z_i})^2$$

$$\sigma_Y^2 = \sum b(i)^2 \times \sigma_{z(i)}^2$$

The percent of contribution from each Z was calculated as follows:

$$\%z_i = \frac{b(i)^2 \times \sigma_{z(i)}^2}{\sigma_Y^2} \times 100\%$$

The contribution of z_1 is 68.241 %
The contribution of z_2 is 14.5954 %
The contribution of z_3 is 17.1277 %
The contribution of z_4 is 0.035818 %

Because of this, we will keep z_1, z_2, and z_3.
The new vector B is:

−3.2903
1.6415
−1.9924

0

The new W matrix:

−0.7808	−0.5914	−0.2012	0
−0.1474	0.4874	−0.8606	0
−0.4293	0.4542	0.3308	0
−0.4293	0.4542	0.3308	0

Next, we calculated the "real" coefficients A based on "artificial" coefficients B_{new}:

$$A_{important} = W_{new} \times B_{new}$$

Our calculated coefficients A are :
1.9992
2.9998
1.4990
1.4990
The coefficient of determination for the new model is 0.99863.

Exercise 2.2: Problem 1

Although the following analysis does not include "all possible combinations of first and second order regressors, it demonstrates the principle of establishing the model configuration using the coefficient of determination

Equation 1 with x_1, x_2, x_3, x_1x_3, x_2^2 has coefficients

$$A_1 = [1.9989 \; 2.9983 \; -0.4002 \; 0.5003 \; 1.0009]$$

Equation 2 with x_1, x_2, x_3, x_1x_3 has coefficients

$$A_2 = [1.9989 \; 2.9983 \; -0.4002 \; 0.5003]$$

Equation 3 with x_1, x_2, x_3, x_2^2 has coefficients

$$A_3 = [1.9989 \; 2.9983 \; -0.4002 \; 1.0009]$$

Equation 4 with x_1, x_2, x_1x_3, x_2^2 has coefficients

$$A_4 = [1.9989 \; 2.9983 \; 0.5003 \; 1.0009]$$

Equation 5 with x_1, x_3, x_1x_3, $x_2{}^2$ has coefficients

$$A_5 = [1.9989 \;\; -0.4002 \;\; 0.5003 \;\; 1.0009]$$

Equation 6 with x_2, x_3, x_1x_3, $x_2{}^2$ has coefficients

$$A_6 = [2.9983 \;\; -0.4002 \;\; 0.5003 \;\; 1.0009]$$

Equation 7 with x_1, x_2, x_1x_3 has coefficients

$$A_7 = [1.9989 \;\; 2.9983 \;\; 0.5003]$$

Equation 8 with x_1, x_2, $x_2{}^2$ has coefficients

$$A_8 = [1.9989 \;\; 2.9983 \;\; 1.0009]$$

For Equations 1–8, the respective natural variance (S_y), error variance (S_e), and coefficient of determination (C_D) values are:

Eqn	S_y	S_e	C_D
1	132.8594	0.0250	0.9998
2	132.8594	28.8738	0.7827
3	132.8594	70.8680	0.4666
4	132.8594	1.5644	0.9882
5	132.8594	31.7197	0.7613
6	132.8594	119.9781	0.0970
7	132.8594	22.0554	0.8340
8	132.8594	53.3903	0.5981

Equation 7, $y = 2x_1 + 3x_2 + 0.5x_1x_3$, seems to be a rational model in terms of complexity and accuracy

Exercise 2.2: Problem 2

The obtained RLSM and "true" parameters are:

RLSM	"True"
7.9989	8.0000
−5.9883	−6.0000
4.9960	5.0000

The coefficient of determination for this model is 0.99994. The plot showing the convergence of the RLSM procedure is shown below. It could be seen that RLSM estimation of constant parameters results in the same parameter values that could be obtained by the LSM.

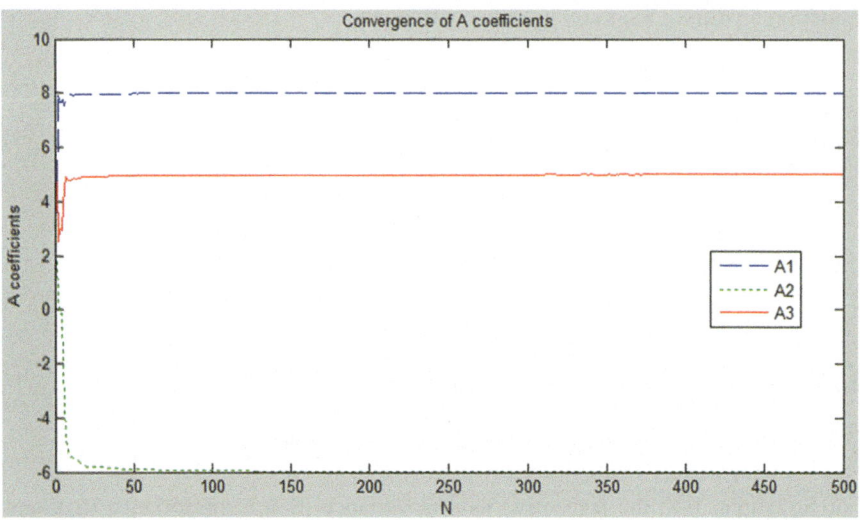

Exercise 2.2: Problem 3

It could be seen that with the forgetting factor of 1. the RLSM procedure does not allow for tracking of drifting "true" parameters. The "final" parameter values are (see the plot below):

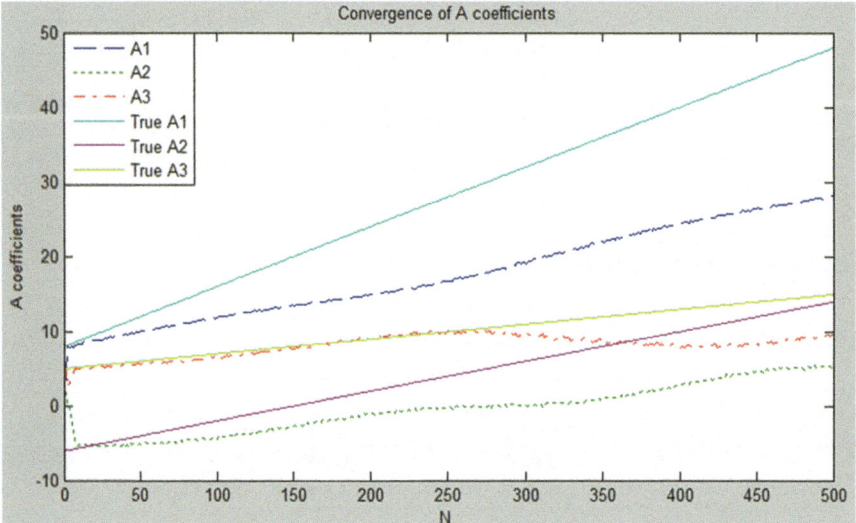

RLSM	"True"
28.0449	48.0000
5.3351	14.0000
9.6195	15.0000

The coefficient of determination for the resultant model is 0.40497, compare with the value of 0.99994 for problem 2. These results are unusable, but justify the use of forgetting factor value of less than 1.

Exercise 2.2: Problem 4

RLSM results with the forgetting factor (Beta) Beta = 0.1 are shown below:

RLSM	True
47.6793	48.0000
13.9316	14.0000
15.4238	15.0000

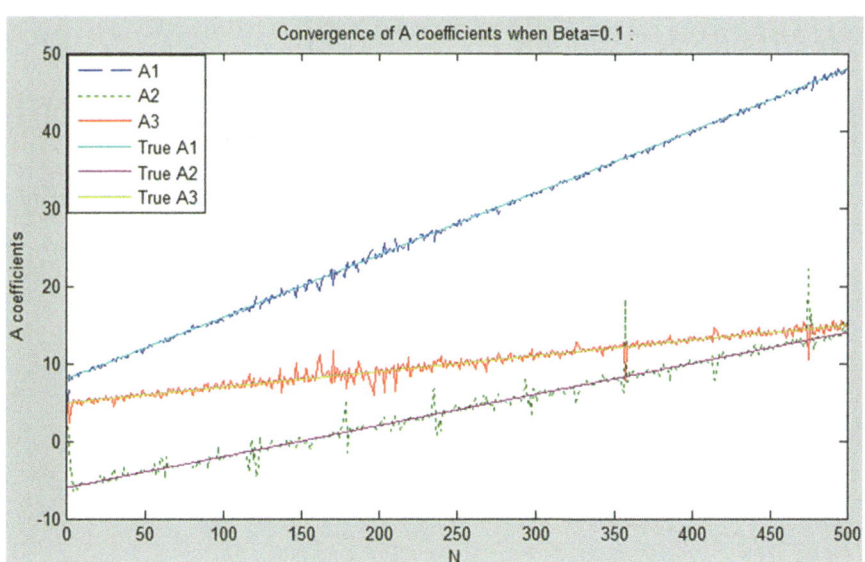

When Beta = 0.2, the results for A are:

RLSM	True
47.6748	48.0000
13.9308	14.0000
15.4287	15.0000

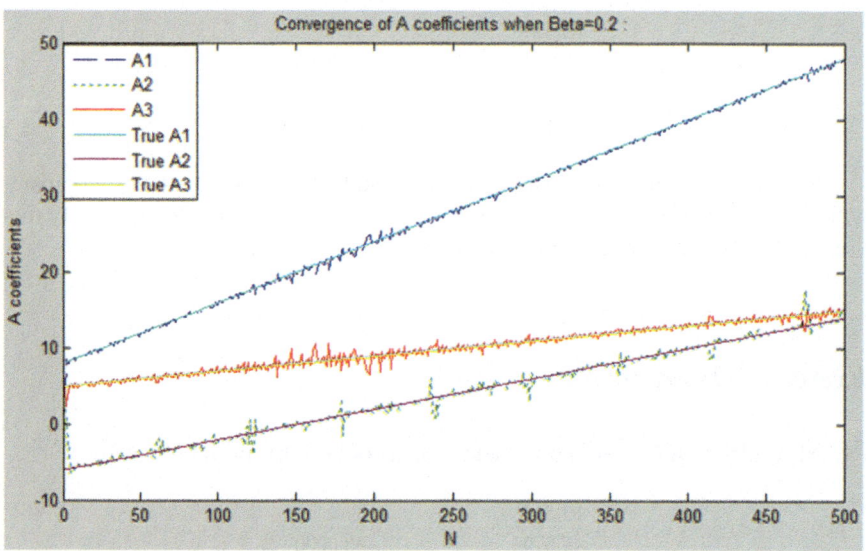

When Beta = 0.3, the results are:

RLSM	True
47.6545	48.0000
13.9228	14.0000
15.4561	15.0000

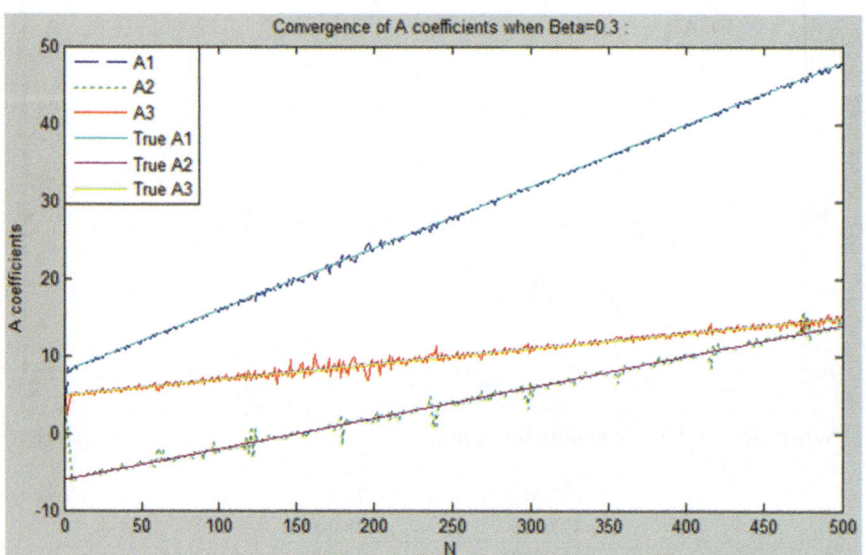

When Beta = 0.4, the converged coefficients of A are:

RLSM	True
47.6172	48.0000
13.9090	14.0000
15.5045	15.0000

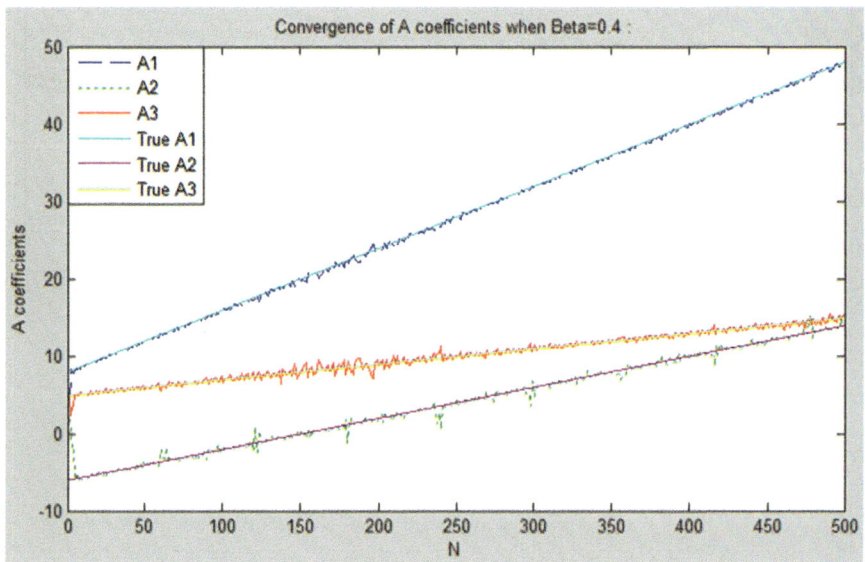

When Beta = 0.5:

RLSM	True
47.5709	48.0000
13.8944	14.0000
15.5529	15.0000

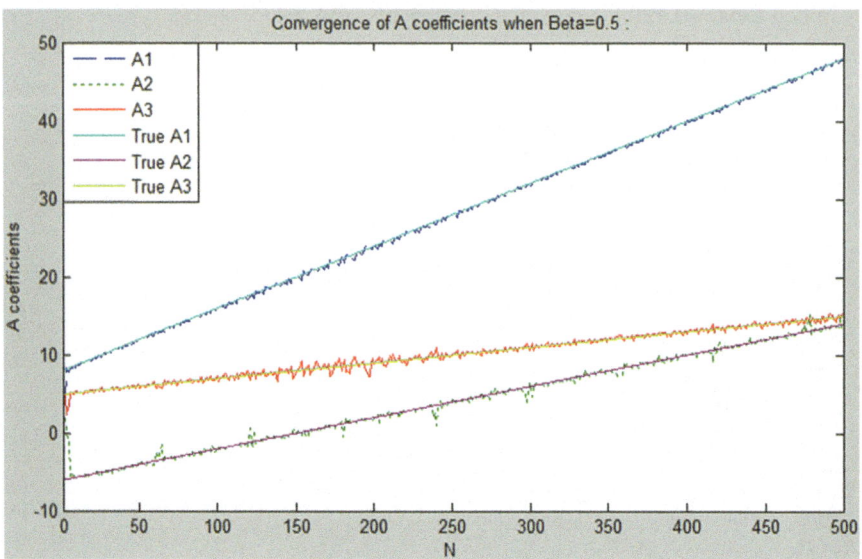

When Beta = 0.6, the converged coefficients of A are:

RLSM	True
47.5318	48.0000
13.8832	14.0000
15.5578	15.0000

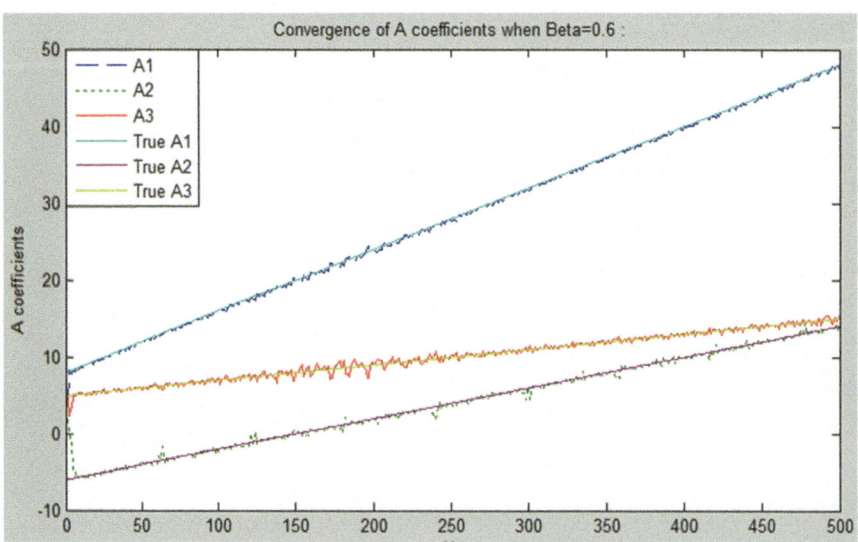

When Beta = 0.9 and Beta = 1. the tracking results are:

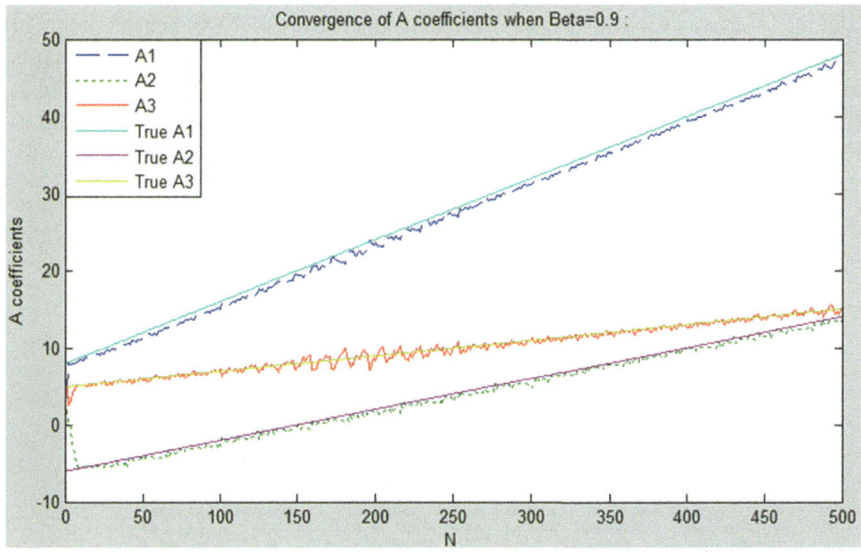

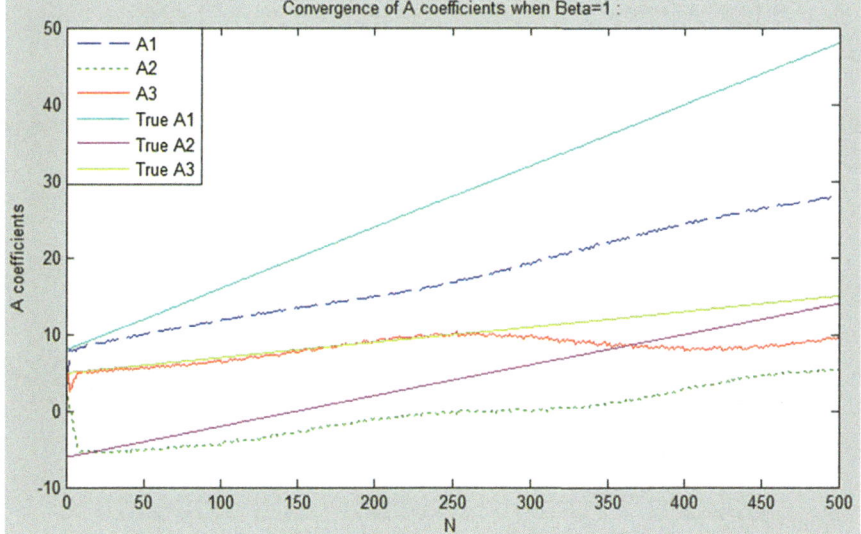

It could be seen that as Beta approaches the value of 1.0 the tracking ability of the RLSM procedure diminishes.

Exercise 2.3: Problem 1

Running program GENERATOR.EXE results in an array of 1000 measurements of a complex process with 15 input variables and a discrete event-type output that is rated as outcome A or outcome B. This data is recorded in file TSMD.DAT. Running program CLUSTER.EXE results in the display of the three most informative subspaces featuring distributions of events A and B in the appropriated subspaces.

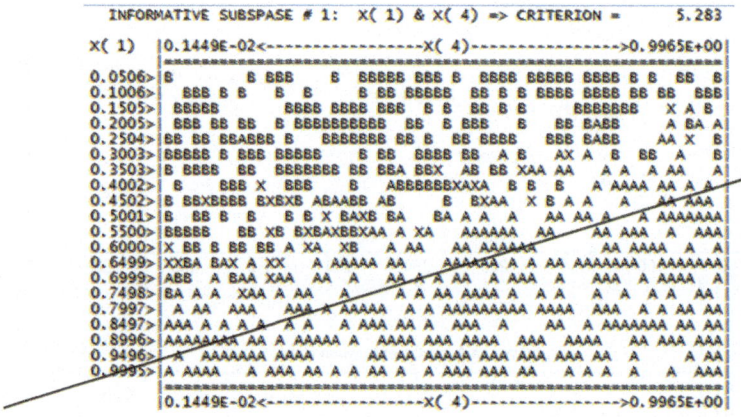

Subspace: X_1 & X_4 - Separation Line is: $X_1 + 1.8139X_4 - 1.4506 = 0$

Subspace: X_1 & X_5 - Separating Line: $X_1 + 2.858X_5 - 2.0003 = 0$

```
INFORMATIVE SUBSPASE # 3:  X( 1) & X(10) => CRITERION =      5.143

X( 1)   |0.5567E-03<----------------X(10)---------------->0.9987E+00|
        |=====================================================================|
0.0506>|BBBBBB   B BB B    BB BBB BB B B B   BBB  BB BBB B      B    BBB|
0.1006>|B    B  BBBB B   BB      B  BB BBB B   BBBB B BBBBBB B BBBB B   |
0.1505>|B B B   BB   B  B A B B   BXBB  B B BBBBBB BB   BB  X B    B B|
0.2005>| BX B BB  BBB    XBB B B   B B   BBB  B BBB  BB BBBX B  X BBBB|
0.2504>|B X  BB BB   BBB   B  BB   BB BBB BB AAB B  B  ABBBB XB   BBX |
0.3003>|B B   B BAB XX   B   B A B BB    BB    BXB A   B    BBB  BBBBBB|
0.3503>|ABBA BABB BBB    X BBX B XXBX B BA  ABXAB   X     BX  BBBBBA  B|
0.4002>|   AB  XBBBBBABBAX AA B      B B AX XA B  AX X XXX ABB BX XA |
0.4502>|  XXBBA AXAA BXBA BAAA A XB BA XX  BBX  B A   AB B  A B BBA|
0.5001>|A XA  A A A BA  AAAA    ABB ABAAA  XBA A BX AB A B   A   XBBB |
0.5500>|XXX ABBA AB B    A AABA BABX A AA  A  AB    BX    AAA XX ABA|
0.6000>|B   A  A AABA A      AA XAAXAAA A  X    B   XAA AA  BABX BA |
0.6499>|AAA  X  A  AXAB AA A AAAAAA  AA   ABAAXA A AA BA  A A   AAA A|
0.6999>|  BA      AA AAX  A A B   A  AA A ABA  AAA AA A A    AA  AA|
0.7498>|  AA    AA  AAA AAAA  A  A  A   A AA AA  A AAX A X  AA A   |
0.7997>| A      AAA   A AAAA  A A A A  AAAA  AAAA AAA AA   A  AAA AA|
0.8497>|  AA A   A  AAA  AA AAAA A AAAAA A  A AAAAAAAA AAA  A AAA A |
0.8996>|A  A AAA  A   A AA AAAAA AA  AA    A  A  A AAAAA AA A AAA  AA|
0.9496>|AAA      AAA  A  AA  AAAAAA A A AAA A AAAAAA AA  A A  AA AA  A|
0.9995>| AAA   A   AAAAAAA AAA A AA A  A  A AA A AAA    AAA A A A  AAA|
        |=====================================================================|
        |0.5567E-03<----------------X(10)---------------->0.9987E+00|
```

Subspace: X_1 & X_{10} - Separating Line: $X_1 + 6.6622X_{10} - 3.6642 = 0$

Events E1–E8 represent location of a measurement point within the domain A or domain B in the above subspaces. Below are the probabilities of these events for outcomes A and B:

```
P[A]=0.563   P[B]=0.437

P[E1/A]=0.037 P[E2/A]=0.048 P[E3/A]=0.108 P[E4/A]=0.119
P[E5/A]=0.140 P[E6/A]=0.138 P[E7/A]=0.176 P[E8/A]=0.234


P[E1/B]=0.284 P[E2/B]=0.296 P[E3/B]=0.140 P[E4/A]=0.126
P[E5/B]=0.057 P[E6/B]=0.060 P[E7/B]=0.016 P[E8/B]=0.021
```

Prediction of the process outcome based on the particular measurement, X(1).

Although vector X(t) has 15 components, our analysis indicates that the values of the following four components are to be considered:

$$X_1(1) = 0.633, \ X_4(1) = 0.814, \ X_5(1) = 0.371, \ X_{10}(1) = 0.363$$

Compute values of functions $\Phi 1$, $\Phi 2$, and $\Phi 3$ for the selected components of vector X(1) and based on these results define the location of the point X(1) as the appropriate event:

$$\Phi 1 = X1(1) + K*X4(1) + Q = 0.633 + 1.8139*0.814 - 1.4506 = 0.6589146$$

$$\Phi2 = X1(1) + K*X5(1) + Q = 0.633 + 2.858*0.371 \; - \; 2.0003$$
$$= -0.306982$$

$$\Phi3 = X1(1) + K*X10(1) + Q = 0.633 + 6.6622*0.363 \; - \; 3.6642$$
$$= -0.6128214$$

The resultant event is E5, then

$$P(A) = 0.563 \quad P(B) = 0.437 \quad P(E5|A) = 0.140 \quad P(E5|B) = 0.057$$

$$P(A|E5) = \frac{P(E5|A)*P(A)}{P(E5|A)*P(A) + P(E5|B)*P(B)} = \frac{0.140*0.563}{0.140*0.563 + 0.057*0.437}$$

$$= .7599$$

$$P(B|E5) = \frac{P(E5|B)*P(B)}{P(E5|B)*P(B) + P(E5|A)*P(A)} = \frac{0.057*0.437}{0.057*0.437 + 0.140*0.563}$$

$$= .2401$$

Consider the next measurement vector X(2) and repeat the above procedure:

$$X_1(2) \; = \; 0.255, \; X_4(2) \; = \; 0.967, \; X_5(2) \; = \; 0.884, \; X_{10}(2) \; = \; 0.067$$
$$\Phi1 = X1(2) + K*X4(2) + Q = 0.255 + 1.8139*0.967 \; - \; 1.4506 = 0.5584413$$
$$\Phi2 = X1(2) + K*X5(2) + Q = 0.255 + 2.858*0.884 \; - \; 2.0003 = 0.781172$$
$$\Phi3 = X1(2) + K*X10(2) + Q = 0.255 + 6.6622*0.067 \; - \; 3.6642$$
$$= -2.9628326$$

That results in Event E7, therefore

$$P(E7|A) = 0.176 \quad P(E7|B) = 0.016 \quad P(A) = \mathbf{0.7599} \quad P(B) = \mathbf{0.2401}$$

$$P(A|E7) = \frac{P(E7|A)*P(A)}{P(E7|A)*P(A) + P(E7|B)*P(B)} = \frac{0.176*0.7599}{0.176*0.7599 + 0.016*0.2401}$$

$$= .9721$$

$$P(B|E7) = \frac{P(E7|B)*P(B)}{P(E7|B)*P(B) + P(E7|A)*P(A)} = \frac{0.016*0.2401}{0.016*0.2401 + 0.176*0.7599}$$

$$= .0279$$

The probability that these two sets of X(t) values yields result B is less than 0.03, while the probability that the outcome would be A is above 0.97. It can be said with much certainty that the outcome associated with these two X(t) sets would be A.

Exercise 2.3: Problem 2

For this problem, a 400×600 random matrix A_0, matrix B_0, and matrix C_0 were generated.

SVD was performed in MATLAB on matrix A_0 to retrieve the first two left and right vectors.

Then, a set of matrices $A_0(k)+noise$ (20 % of the original magnitude used to generate matrix A), k=1,2,3,10 and two-coordinate points $\{W_1(k),W_2(k)\}$ were defined by multiplication:

$$W(k)_1 = L_{A1}{}^T \times A(k) \times R_{A1}$$
$$W(k)_2 = L_{A2}{}^T \times A(k) \times R_{A2}$$

k=1,2,...10

This process was repeated still using the left and right vectors of the original matrix A_0, but instead of $A(k)$ matrices $B(k)$ and $C(k)$, generated by adding noise to B_0 and C_0, were used, and a sequence of points $\{W_1(k), W_2(k)\}$, k=10+1, 10+2,...20, 20+1, 20+2,...,30 were established.

All of the points were plotted on a W_1–W_2 plane. As it can be seen, this SVD-based procedure results in the clustering pattern revealing in spite of noise the three classes of matrices originated from matrix A_0, B_0, C_0.

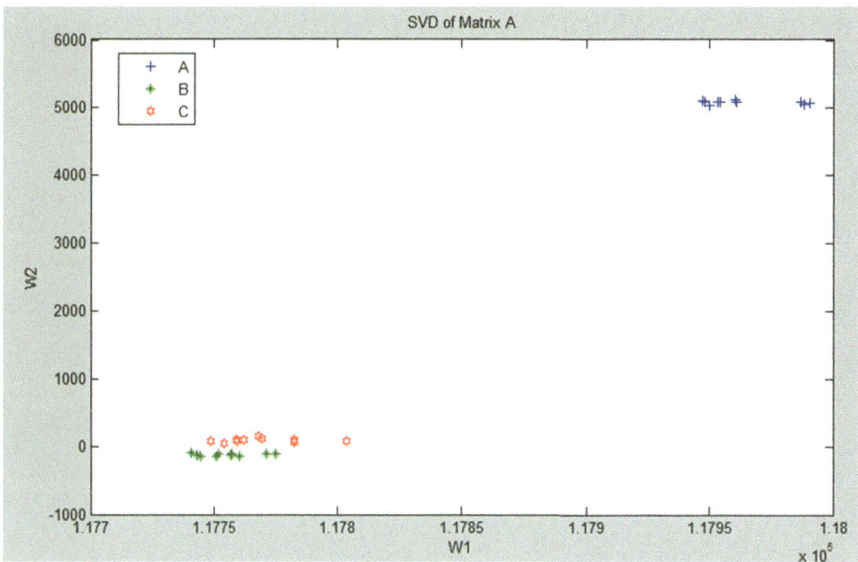

Indeed, the SVD could be used as a tool for classification of large groups of data sets.

Bibliography

1. Sam Kash Kachigan, Multivariate Statistical Analysis: A Conceptual Introduction, 2nd Edition, ISBN-13: 978-0942154917
2. Alvin C. Rencher, William F. Christensen, Methods of Multivariate Analysis 3rd Edition, Wiley, ISBN-13: 978-0470178966
3. Astrom, K.J. and Wittenmark, B., Adaptive Control, Addison-Wesley Publishing Company.
4. V. Skormin, V. Gorodetski, L. Popyack, "Data Mining Technology for Failure Prognostics of Avionics," *IEEE Trans. On Aerospace and Electronic Systems* 38, no. 2 (Apr 2002): 388–402.
5. A. Tarakanov and V. Skormin, Immunocomputing: Principles and Applications, Springer, 2002

Chapter 3
Computer Control of Manufacturing Processes

The availability of a mathematical description of a manufacturing process provides a quantitative basis for process control that is understood as maintaining the desired status of the process in spite of many external and internal disturbance effects. The desired status is defined by the set points, also known as reference signals that represent the desired numerical values of controlled variables of the process. The control task implies that the difference between the actual and desired process variables (error) is determined, and on the basis of this difference and according to the control law, the control efforts are defined and applied to the process. The manner in which the control effort drives the system from its actual state to the desired state, and the allowable discrepancy between the actual and the desired states are quite important. They are dependent on the control law and are explicitly defined by the design specifications along with the discrete time step (or clock frequency). Modern computer systems facilitate every function of the control task; process monitoring, specification of the set points, extraction of the errors, implementation of the control law, and the application of control efforts.

3.1 S- and Z-Domain Transfer Functions

An s-domain transfer function, defined as a "Laplace transform of the output signal over Laplace transform of the input signal under zero initial conditions" presents the most common technique for the mathematical description of linear, dynamic, single-input-single-output systems. In the case of a multi-input-multi-output linear dynamic system, transfer functions of particular channels could be easily assembled into a transfer matrix. Transfer functions enable us to address the rigors of linear differential equations through simple algebra. Control engineers commonly use s-domain transfer functions for the analysis and design of continuous-time control systems. The situation changes when a discrete-time control system is to be developed, which is a very typical case in our computer-dominated environment.

© Springer International Publishing Switzerland 2016
V.A. Skormin, *Introduction to Process Control*, Springer Texts
in Business and Economics, DOI 10.1007/978-3-319-42258-9_3

Although discrete-time control synthesis requires the use of z-domain transfer functions, the material of this chapter allows students to utilize their experience with s-domain transfer functions to the fullest.

A z-domain transfer function, defined as a "Z-transform of the output signal over Z-transform of the input signal under zero initial conditions", is used for mathematical description of linear, dynamic, single-input-single-output systems operated by a discrete-time controller. The output of such a controller is a "number sequence" i.e. a sequence of real numbers, $u^*(k)$, $k = 1,2,3,\ldots$, generated at the clock frequency, f^{CL}, of the computer implementing the control law. The purpose of the asterisk is to remind the reader that this number should be represented by an analog signal, $w(t)$, known as a "pulse train"

$$w(t) = \begin{cases} u^*(k) & \text{if } t = kT \\ 0 & \text{if } (k-1)T < t < kT \end{cases}$$

where $T = \frac{1}{f^{CL}}$ is the time step. A "number sequence" simply has no power to drive the control plant, but it serves as the input to many different devices labeled as the *zero-order-hold* that convert these periodically generated real numbers into a "staircase"-type control effort

$$u^*(k) \quad \text{if } (k)T \le t < (k+1)T$$

Note that $\bar{u}^*(k)$ is not equal to $u^*(k)$; in actuality $\bar{u}^*(k)$ is the approximation of $u^*(k)$ by the integer number of discretization steps Δu, so that $u^*(k) \approx \Delta u \cdot \bar{u}^*(k)$. It looks like this representation results in a special round-off error that is being introduced in the system at each time step. This is true, but the error process is dominated by clock frequency f^{CL} that is normally chosen well beyond the system bandwidth.

The utilization of a zero-order-hold imposes some requirements on the mathematical description of the controlled plant in the discrete-time domain. Note that the direct z-domain equivalent of an s-domain transfer function obtained for a particular time step T

$$G(s) \xrightarrow{T} G(z) = Z\{G(s)\}$$

is suitable only when the process is driven by a pulse train signal. In the case of a "controlled plant driven through a zero-order-hold" the z-domain transfer function of the plant is defined as

$$G(z) = (1 - z^{-1})Z\left\{\frac{1}{s}G(s)\right\}$$

This conversion is routinely performed using a software tool and specifying the required time step T and the ZOH conversion option:

$$G(s) \xrightarrow[\text{ZOH}]{\text{T}} G(z) = \left(1 - z^{-1}\right) Z \left\{ \frac{1}{s} G(s) \right\}$$

Speaking of this conversion, it should be mentioned that $G(z)$ is expected to be of the same order as $G(s)$, and $G(z)$ may have zeros even if $G(s)$ does not. In the future, we will always refer to z-domain transfer functions intended for the zero-order-hold applications.

Consider a z-domain transfer function $G(z) = \frac{N(z)}{D(z)} = \frac{Y(z)}{U(z)}$ where

$$N(z) = b_m z^m + b_{m-1} z^{m-1} + \ldots + b_1 z + b_0 ,$$
$$D(z) = z^n + a_{n-1} z^{n-1} + a_{n-2} z^{n-2} + \ldots + a_1 z + a_0$$

are numerator and denominator of the transfer function, $n \geq m$, and $Y(z)$ and $U(z)$ are Z-transforms of the output and input signals,

$$\frac{Y(z)}{U(z)} = \frac{b_m z^m + b_{m-1} z^{m-1} + \ldots + b_1 z + b_0}{z^n + a_{n-1} z^{n-1} + a_{n-2} z^{n-2} + \ldots + a_1 z + a_0}$$

The following is the discrete-time equivalent of the above relationship,

$$y(k+n) = -a_{n-1} y(k+n-1) - a_{n-2} y(k+n-2) - \ldots - a_1 y(k+1) - a_0 y(k)$$
$$+ b_m u(k+m) + b_{m-1} u(k+m-1) + \ldots + b_1 u(k+1) + b_0 u(k)$$

Example 3.1 Given transfer function of a controlled process, $G(s) = \frac{2s^2+5s+3}{s^3+5s^2+10s+8}$, obtain its discrete-time mathematical description assuming that this process will be driven through a zero-order-hold and the time step $T = 0.02$ s. First, let us convert $G(s)$ into $G(z)$ for the required time step and using the ZOH option:

$$\frac{2s^2 + 5s + 3}{s^3 + 5s^2 + 10s + 8} \xrightarrow[\text{ZOH}]{\text{T}=.02} G(z) = \frac{.039z^2 - .076z + .037}{z^3 - 2.091z^2 + 2.806z - .905}$$

Assume that $u(k)$ and $y(k)$ are the discrete-time input and output signals, then the discrete-time description of the $y(k)$–$u(k)$ relationship could be obtained as follows:

$$\frac{Y(z)}{U(z)} = \frac{.039z^2 - .076z + .037}{z^3 - 2.091z^2 + 2.806z - .905} \rightarrow Y(z)\left(z^3 - 2.091z^2 + 2.806z - .905\right)$$
$$= U(z)(.039z^2 - .076z + .037)$$

and

$$y(k+3) = 2.091y(k+2) - 2.806y(k+1) + .905y(k) + .039u(k+2)$$
$$- .076u(k+1) + .037u(k)$$

Note that although the above expression properly represents the dynamic relationship between variables y(k) and u(k), it is not particularly usable for computing values of the output, y(k), given values of the input, u(k). Indeed, if k represents the current time, the above expression deals with the future, presumably unknown, values of the variables. Therefore, it must be converted to a *recursive formula* relating the current value of the output, y(k), to the already known values of the input and output. The recursive formula could be obtained by dividing the numerator and denominator of the z-domain transfer function by the highest power of z and making the following transformations:

$$\frac{Y(z)}{U(z)} = \frac{.039z^2 - .076z + .037}{z^3 - 2.091z^2 + 2.806z - .905} = \frac{.039z^{-1} - .076z^{-2} + .037z^{-3}}{1 - 2.091z^{-1} + 2.806z^{-2} - .905z^{-3}} \longrightarrow$$

$$Y(z)(1 - 2.091z^{-1} + 2.806z^{-2} - .905z^{-3}) = U(z)(.039z^{-1} - .076z^{-2} + .037z^{-3}), \text{ and}$$

$$y(k) = 2.091y(k-1) - 2.806y(k-2) + .905y(k-3) + .039u(k-1) - .076u(k-2)$$
$$+ .037u(k-3)$$

thus enabling us to compute the output on the basis of already existing input/output observations.

The following computer code could be written to perform this task (assuming zero initial conditions):

$$\text{Initialization} \begin{cases} \texttt{y3old} = 0 \\ \texttt{y2old} = 0 \\ \texttt{y1old} = 0 \\ \texttt{u3old} = 0 \\ \texttt{u2old} = 0 \\ \texttt{u1old} = 0 \end{cases}$$

```
Beginning of the loop
```

$$\text{Loop} \begin{cases} \texttt{Input (...,...) u} \\ \texttt{y} = 2.091\texttt{*y1old} - 2.806\texttt{*y2old} + .905\texttt{*y3old} + \\ \quad .039\texttt{*u1old} - .076\texttt{*u2old} + .037\texttt{*u3old} \\ \texttt{y3old} = \texttt{y2old} \\ \texttt{y2old} = \texttt{y1old} \\ \texttt{y1old} = \texttt{y} \\ \texttt{u3old} = \texttt{u2old} \\ \texttt{u2old} = \texttt{u1old} \\ \texttt{u1old} = \texttt{u} \\ \texttt{Output (...,...) y} \end{cases}$$

```
Return to the beginning of the loop
```

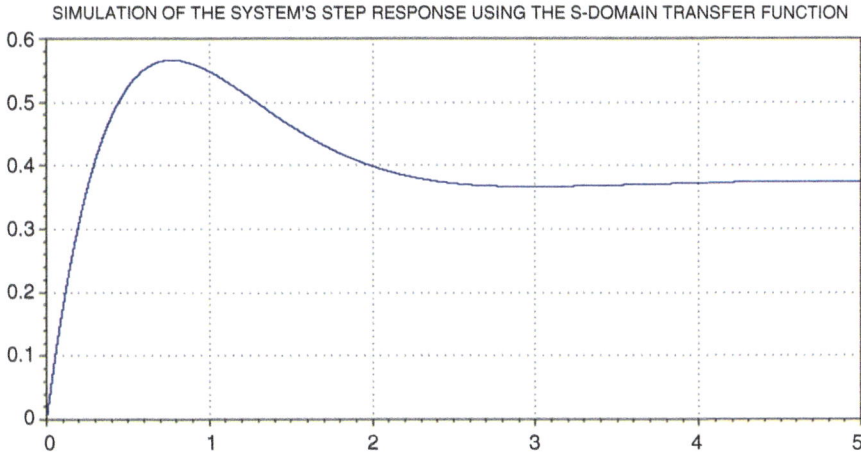

Fig. 3.1 s-domain transfer function step response

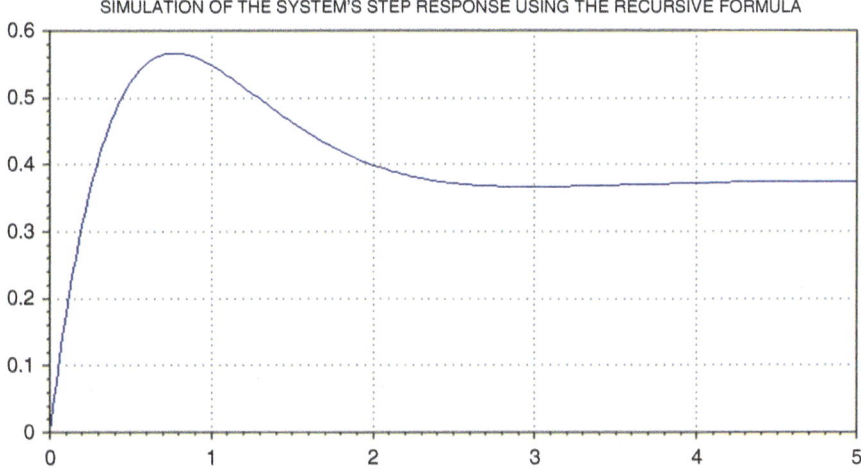

Fig. 3.2 Recursive formula representation step response

The following Figs. 3.1 and 3.2 are the step responses of an analog system described by transfer function $G(s) = \frac{2s^2+5s+3}{s^3+5s^2+10s+8}$ and the values of $y(k)$, $k = 0,1,2,3,\ldots$ obtained on the basis of the recursive formula

$$y(k) = 2.091y(k-1) - 2.806y(k-2) + .905y(k-3) + .039u(k-1) - .076u(k-2) + .037u(k-3)$$

assuming that $u(k) \equiv 1$. One can realize that the responses completely match.

3.2   Mathematical Modeling of Dynamic Systems

It has been agreed that a z-domain transfer function is the most attractive form of mathematical description of a dynamic system facilitating the controls related tasks. We will consider the development of such transfer functions using the input/output data, $u(k)$, $y(k)$, $k = 1,2,3,\ldots$. Unlike a static model, relating immediate values of input/output variables, i.e. $y(t) = \varphi[u(t)]$, a dynamic model relates the immediate value of the output variable to the immediate value of the input variable, to the previous values of the input variables, and to the previous values of the output variables, i.e. $y(t) = \varphi[u(t), u(t-\tau), y(t-\tau)]$. In the case of a discrete-time model of a dynamic system, relating its discrete-time values, a dynamic model can be represented as

$$y(k+n) = \varphi[y(k+n-1), y(k+n-2), \ldots, y(k), u(k+n), u(k+m-1),$$
$$u(k+m-2), \ldots, u(k)], \; k = 1,2,3, \ldots.$$

An expression of this type can also be called a *recursive formula*. In the case of a linear model the above expression can be written as

$$y(k+n) = \sum_{j=1}^{n} a_{n-j} y(k+n-j) + \sum_{j=0}^{m} b_{m-j} u(k+m-j), \; k = 1,2,3, \ldots.$$

Having a recursive formula allows for straightforward definition of the z-domain transfer function. Taking the Z-transform of the left-hand and right-hand sides of the above expression and assuming zero initial conditions results in

$$z^n Y(z) = \sum_{j=1}^{n} a_{n-j} z^{n-j} Y(z) + \sum_{j=0}^{m} b_{m-j} z^{m-j} U(z)$$

Further transformation of this expression leads to the definition of a z-domain transfer function:

$$Y(z)\left(z^n - \sum_{j=1}^{n} a_{n-j} z^{n-j}\right) = U(z) \sum_{j=0}^{m} b_{m-j} z^{m-j}$$

$$\frac{Y(z)}{U(z)} = \frac{\displaystyle\sum_{j=0}^{m} b_{m-j} z^{m-j}}{1 - \displaystyle\sum_{j=1}^{n} a_{n-j} z^{n-j}} = \frac{b_m z^m + b_{m-1} z^{m-1} + b_{m-2} z^{m-2} + \ldots + b_0}{z^n - a_{n-1} z^{n-1} - a_{n-2} z^{n-2} - \ldots - a_0} = G(z)$$

In many practical situations the expression for the linear dynamic model is known in the form of a z-domain transfer function G(z) or a recursive formula, but its parameters (coefficients), a_j and b_j, are unknown. Then the modeling problem is reduced to the estimation of the unknown parameters on the basis of the available input/output measurements, u(k), y(k), k = 1,2,3,..., N.

In order to utilize the familiar LSM, consider the recursive formula again:

$$y(k+n) = \sum_{j=1}^{n} a_{n-j} y(k+n-j) + \sum_{j=0}^{m} b_{m-j} u(k+m-j)$$

Introduce a set of new variables:

$$v(k) = y(k+n)$$
$$x_1(k) = -y(k+n-1) \quad x_{n+1}(k) = u(k+m)$$
$$x_2(k) = -y(k+n-2) \quad x_{n+2}(k) = u(k+m-1)$$
$$\ldots \quad\quad\quad\quad \ldots$$
$$x_{n-1}(k) = -y(k+1) \quad x_{n+m}(k) = u(k+1)$$
$$x_n(k) = -y(k) \quad\quad x_{n+m+1}(k) = u(k)$$

Since we are not equipped for dealing with negative discrete-time index values, and n > m, assume that k = 1,2,..., N−n. Introduce vector

$$x(k) = [x_1(k), x_2(k), \ldots, x_n(k), \ldots, x_{n+1}(k)]^T, \ k = 1, 2, 3, \ldots N-n$$

and organize all values of x(k) into the array X (realize that transposed vectors x (k) serve as rows in this array and it contains only N−n rows). Measurements v(k), k = 1, 2,..., N−n could be arranged in array V that also has N−n rows. Now our original recursive formula could be represented as

$$v(k) = x(k)^T C,$$

where

$$x(k) = \begin{bmatrix} x_1(k) \\ x_2(k) \\ \ldots \\ x_{n+m}(k) \\ x_{n+m+1}(k) \end{bmatrix} \quad \text{and} \quad C = \begin{bmatrix} c_1 \\ c_2 \\ \ldots \\ c_{n+m} \\ c_{n+m+1} \end{bmatrix} = \begin{bmatrix} -a_{n-1} \\ -a_{n-2} \\ \ldots \\ -a_0 \\ b_m \\ \ldots \\ b_0 \end{bmatrix}.$$

It is understood that known measurements $u(k)$, $y(k)$, $k = 1,2,3,\ldots,N$ result in the known arrays X_N and V_N, and unknown coefficients C can be estimated using the LSM approach, i.e.,

$$C = \left(X^T X\right)^{-1}\left(X^T V\right)$$

and then interpreted as parameters of the z-domain transfer function

$$G(z) = \frac{Y(z)}{U(z)} = \frac{b_m z^m + b_{m-1} z^{m-1} + b_{m-2} z^{m-2} + \ldots + b_0}{z^n + a_{n-1} z^{n-1} + a_{n-2} z^{n-2} + \ldots + a_0}$$

Moreover, the RLSM could be utilized for the same estimation task which is even more desirable since it assures that the most recent changes in the controlled plant are reflected by the model. Note that all techniques for model validation and confidence analysis of model parameters are applicable in this case.

It is good to know that LSM/RLSM –based parameter estimation of a dynamic model results in the coefficients of a z-domain transfer function intended for the zero-order-hold applications.

Example 3.2 Estimation of parameters of a z-domain transfer function

$$G(z) = \frac{b_1 z + b_0}{z^2 + a_1 z + a_0}$$

of a controlled plant utilizing input/output measurement data.

First, realize that computationally this problem can be reduced to the estimation of coefficients of the following regression equation:

$$v(k) = c_1 x_1(k) + c_2 x_2(k) + c_3 x_3(k) + c_4 x_4(k)$$

where $v(k) = y(k+2)$, $x_1(k) = -y(k+1)$, $x_2(k) = -y(k)$, $x_3(k) = u(k+1)$, and $x_4(k) = u(k)$.

The following are the measurement data for $u(k)$, $y(k)$, $k = 1,2,\ldots$, and arrays X and V:

$$
\begin{bmatrix} k \\ 1 \\ 2 \\ 3 \\ 4 \\ 5 \\ 6 \\ 7 \\ 8 \\ 9 \\ 10 \\ 11 \\ 12 \\ 13 \\ 14 \\ 15 \\ 16 \end{bmatrix}
\begin{bmatrix} u \\ -1.723 \\ 1.871 \\ 16.15 \\ -1.768 \\ 6.531 \\ -5.464 \\ 1.941 \\ 9.257 \\ 12.040 \\ 15.310 \\ -13.56 \\ 0.515 \\ 10.20 \\ -12.26 \\ 7.085 \\ 8.717 \end{bmatrix}
\begin{bmatrix} y \\ 0.000 \\ -0.101 \\ 0.014 \\ 0.959 \\ 0.809 \\ 1.152 \\ 0.775 \\ 0.849 \\ 1.347 \\ 1.984 \\ 2.780 \\ 1.845 \\ 1.777 \\ 2.278 \\ 1.438 \\ 1.769 \end{bmatrix}
X =
\begin{bmatrix}
0.101 & 0.000 & 1.871 & -1.723 \\
-0.014 & 0.101 & 16.150 & 1.871 \\
-0.959 & -0.014 & -1.768 & 16.15 \\
-0.809 & -0.959 & 6.531 & -1.768 \\
1.152 & -0.809 & -5.464 & 6.531 \\
-0.775 & -1.152 & 1.941 & -5.464 \\
-0.849 & -0.775 & 9.257 & 1.941 \\
-1.347 & -0.849 & 12.04 & 9.257 \\
-1.984 & -1.347 & 15.31 & 12.04 \\
-2.780 & -1.984 & -13.56 & 15.31 \\
-1.845 & -2.780 & 0.515 & -13.56 \\
-1.777 & -1.845 & 10.20 & 0.515 \\
-2.278 & -1.777 & -12.26 & 10.20 \\
-1.438 & -2.278 & 7.085 & -12.26
\end{bmatrix} ,
$$

$$
V =
\begin{bmatrix}
0.014 \\
0.959 \\
0.809 \\
1.152 \\
0.775 \\
0.849 \\
1.347 \\
1.984 \\
2.780 \\
1.845 \\
1.777 \\
2.278 \\
1.438 \\
1.769
\end{bmatrix}
$$

then $(X^T X)^{-1}(X^T V) = \begin{bmatrix} 0.0023 \\ -0.892 \\ 0.0589 \\ 0.0568 \end{bmatrix} = C$ and the resultant z-domain transfer function

is

$$
G(z) = \frac{Y(z)}{U(z)} = \frac{0.0589 + 0.0568z^{-1}}{1 - 0.0023z^{-1} + 0.892z^{-2}} = \frac{(0.0589z + 0568)z}{z^2 - 0.0023z + 0.892}
$$

Example 3.3 Investigation of the effect of measurement noise on parameter estimation.

Assume that a continuous-time controlled plant is described by the following transfer function:

$$G(s) = \frac{5s^2 + 6s + 1}{(s^2 + 5s + 10)(s + 11)}$$

The conversion of this transfer function in the z-domain for the time step of 0.01 s and the ZOH option yields

$$G(z) = \frac{0.04644(z^2 - 1.988z + 0.9881)}{(z^2 - 1.95z + 0.9512)(z - 0.8958)} = \frac{0.046z^2 - 0.092z + 0.046}{z^3 - 2.846z^2 + 2.698z - 0.852}$$

A software tool has been used to simulate the response of the continuous-time system to a random signal that was sampled every 0.01 s and the resultant input/output data, y(k), u(k), k = 1,2,..., 1500, was placed in a file. On the basis of this data, a 1500×6 array X and a 1500×1 array V were obtained. The following are covariance matrices K_{XX} and K_{XV}:

$$K_{XX} = \begin{bmatrix} 1.007 & 0.884 & 0.778 & -0.179 & -5.024 & -4.239 \\ 0.884 & 1.007 & 0.883 & -0.283 & -0.175 & -5.020 \\ 0.778 & 0.883 & 1.006 & -0.289 & -0.280 & -0.173 \\ -0.179 & -0.283 & -0.289 & 105.6 & -0.759 & 1.309 \\ -5.024 & -0.175 & -0.280 & -0.759 & 105.6 & -0.790 \\ -4.239 & -5.020 & -0.173 & 1.309 & -0.790 & 105.6 \end{bmatrix},$$

$$K_{XV} = \begin{bmatrix} -0.884 \\ -0.778 \\ -0.684 \\ 5.029 \\ 4.244 \\ 3.644 \end{bmatrix}$$

Application of the LSM procedure results in the estimation of the parameters of the regression equation

$$C = K_{XX}^{-1} * K_{XV} = \begin{bmatrix} -2.846 \\ 2.698 \\ -0.852 \\ 0.046 \\ -0.092 \\ 0.046 \end{bmatrix}$$

that can easily be recognized as particular parameters of the z-domain transfer function.

Now let us introduce "measurement noise" with the standard deviation of 0.0002 in the output channel by adding appropriately chosen random signal to the [y(k), k = 1,2,...] data.

This results in a significant change in the parameters C due to the fact that in the regression equation, representing the dynamic model some of the inputs (regressors) are nothing but the output variable shifted in time. Consequently, the "noise in the output" becomes the "input noise"

$$C = K_{XX}^{-1}*K_{XV} = \begin{bmatrix} -1.834 \\ 0.811 \\ 0.027 \\ 0.046 \\ -0.045 \\ -0.001 \end{bmatrix}$$

At the same time, knowing the variance of this noise can be quite fruitful for improving the estimates. Let us approximate the covariance matrix of the noise as follows:

$$K_{noise} = \begin{bmatrix} \sigma_n^2 & 0 & 0 & 0 & 0 & 0 \\ 0 & \sigma_n^2 & 0 & 0 & 0 & 0 \\ 0 & 0 & \sigma_n^2 & 0 & 0 & 0 \\ 0 & 0 & 0 & 0 & 0 & 0 \\ 0 & 0 & 0 & 0 & 0 & 0 \\ 0 & 0 & 0 & 0 & 0 & 0 \end{bmatrix} = \begin{bmatrix} 4e-8 & 0 & 0 & 0 & 0 & 0 \\ 0 & 4e-8 & 0 & 0 & 0 & 0 \\ 0 & 0 & 4e-8 & 0 & 0 & 0 \\ 0 & 0 & 0 & 0 & 0 & 0 \\ 0 & 0 & 0 & 0 & 0 & 0 \\ 0 & 0 & 0 & 0 & 0 & 0 \end{bmatrix}$$

where $\sigma_n^2 = 4e\text{-}8$ is the variance of the noise in the output channel (note that only first three regressors are defined as shifted output y(k)). Now the improved parameter estimates can be obtained as

$$C = (K_{XX} - K_{noise})^{-1}*K_{XV} = \begin{bmatrix} -2.823 \\ 2.655 \\ -0.832 \\ 0.046 \\ -0.091 \\ 0.045 \end{bmatrix}$$

Model validation includes,

(1) Computation of the modeling error: $E = V - X*C$,
(2) Computation of the variance of the modeling error: $\sigma_{ERR}^2 = \frac{1}{1500} E^T E = 1e\text{-}6$
(3) Computation of the natural variance of variable y (using the CC function):
$\sigma_Y^2 = \text{var}(V) = 0.95$
(4) Computation of the coefficient of determination η, which in our case is very close to 1. It can be concluded that a high quality model was obtained.

3.3   Control System with an Output Feedback Controller

Since most realistic manufacturing processes are continuous-time processes, consider its original mathematical description in the form of an s-domain transfer function G(s). To simplify the presentation, assume that transfer function G(s) does not have right-hand-side zeros. It is quite common that the exact definition of the transfer function is unknown, and the system designer must rely on its discrete-time equivalent G(z) obtained by regression analysis. Recall that G(z) corresponds to the chosen discrete time step T and reflects the most realistic scheme of discrete-time control: the plant is driven through a zero-order-hold. Assume that the design specifications are given by the system bandwidth, ω^{BW}, and the disturbance rejection, δ db.

First, let us present the general description of the design methodology.

Assume that the transfer function of the controlled plant is given in the form $G_P(z) = \frac{N_P(z)}{D_P(z)}$ where $N_P(z)$ and $D_P(z)$ are m-th order polynomial numerator and n-th polynomial denominator, and $n \geq m$. Similarly, the model transfer function, originally defined in the s-domain to comply with the system design specifications and then converted into the z-domain (for the appropriate time step and the ZOH option), is $G_M(z) = \frac{N_M(z)}{D_M(z)}$ where $N_M(z)$ and $D_M(z)$ are n-1-th order polynomial numerator and n-th polynomial denominator. Next, filter $Q(z) = \frac{1}{N_P(z)}$ is to be introduced in the input of the plant and a feedback H(z), where H(z) is a n−1 order polynomial must be introduced forming the system configuration in Fig. 3.3.

It could be seen that the closed-loop transfer function of this system is

$$G_{CL}(z) = \frac{Q(z)\frac{N_P(z)}{D_P(z)}}{1 + H(z)Q(z)\frac{N_P(z)}{D_P(z)}} = \frac{\frac{1}{N_P(z)}\frac{N_P(z)}{D_P(z)}}{1 + H(z)\frac{1}{N_P(z)}\frac{N_P(z)}{D_P(z)}} = \frac{1}{D_P(z) + H(z)}$$

It is necessary to ensure that the closed-loop transfer function $G_{CL}(z)$ can be modified to become equal to the model transfer function. First, the characteristic polynomial of the closed-loop system must be equal to the characteristic polynomial of the model transfer function, i.e.

$$D_P(z) + H(z) = D_M(z)$$

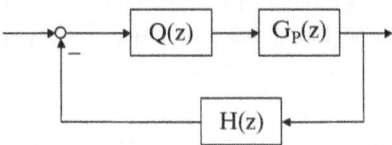

Fig. 3.3 System with filter and feedback controller

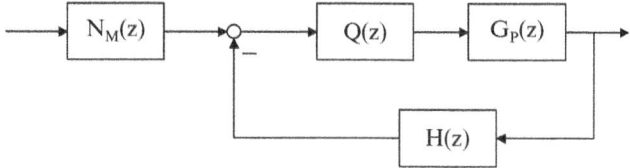

Fig. 3.4 System with filter, feedback controller, and input pre-filter

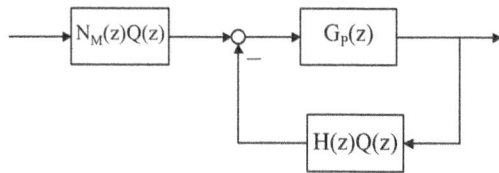

Fig. 3.5 Simplified system from Fig. 3.4

that results in the straight-forward definition of the polynomial H(z):

$$H(z) = D_M(z) - D_P(z)$$

Second, the zeros of the closed-loop transfer function must be equal to the zeros of the model transfer function, that can be achieved by placing a polynomial filter $N_M(z)$ in the reference channel of the above system. This setup is shown in Fig. 3.4.

Finally, the above block diagram in Fig. 3.4 should be transformed resulting in the configuration of the control system as shown below in Fig. 3.5.

Note that the presented design procedure does not address the steady-state error requirements that could be a part of design specifications. These requirements could be satisfied by the manipulation of the non-dominant poles of the model transfer function and/or introduction of an integrator in the control loop as will be demonstrated by the numerical examples below.

The design methodology could be best illustrated by the following numerical example.

Example 3.4 Given the transfer function of a fourth order controlled plant, established on the basis of input/output measurement data (time step of 0.01 s):

$$G_P(z) = \frac{.01(z^3 - 2.94z^2 + 2.88z - .94)}{z^4 - 3.98z^3 + 5.94z^2 - 3.94z + 0.98}$$

The design specifications call for the system bandwidth $\omega^{BW} = 10$ rad/s and the disturbance rejection $\delta = 15$ db.

Recall that $G_P(z)$ is consistent with the time step of 0.01 s and the ZOH case.

First, let us introduce an s-domain model transfer function, $G_M(s)$, of the same order as the controlled plant, representing a system with the desired bandwidth of

10 rad/s. Recall that the frequency response of this system must have the magnitude of zero db within the frequency range $0 < \omega \le \omega^{BW}$ and then should exhibit a magnitude drop of 20 db/dec or more. For this purpose it is recommended to choose a system with a dominant second order term with the damping ratio of 0.8 and the natural frequency of 10 rad/s. The non-dominant part of the characteristic polynomial may have two first order terms. For low overshoot of the system step response it is recommended not to have zeros, and the constant gain of the transfer function shall be chosen to assure that $G^{CL}(0) = 1$. One can realize that this transfer function represents the desired closed-loop dynamics of the reference channel of the system to be designed:

$$
\begin{aligned}
G^{CL}(s) &= \frac{200000}{(s^2 + 2 \cdot 0.8 \cdot 10 \cdot s + 10^2)(s + 40)(s + 50)} \\
&= \frac{200000}{(s^2 + 16 \cdot s + 100)(s + 40)(s + 50)}
\end{aligned}
$$

Obtain the z-domain equivalent of this transfer function for the ZOH option and the time step of 0.01 s (a computer tool is recommended for this task):

$$
G^{CL}(z) = \frac{6.76e - 05(z^3 + 8.94z^2 + 7.24z + .53)}{z^4 - 3.12z^3 + 3.61z^2 - 1.84z + .35}
$$

Note that although the s-domain transfer function $G^{CL}(s)$ does not have zeros, its z-domain equivalent, $G^{CL}(z)$ has zeros.

Assume that filter

$$
Q(z) = \frac{1}{.01(z^3 - 2.94z^2 + 2.88z - .94)} = \frac{100}{z^3 - 2.94z^2 + 2.88z - .94}
$$

is placed in the input of the controlled plant and a polynomial feedback,

$$
H(z) = h_3 z^3 + h_2 z^2 + h_1 z + h_0
$$

is introduced. It could be seen that the overall transfer function of this system is

$$
G^{OV}(z) = \frac{1}{z^4 + (h_3 - 3.98)z^3 + (h_2 + 5.94)z^2 + (h_1 - 3.94)z + h_0 + .98}
$$

and its characteristic polynomial must be equal to the denominator of the desired closed-loop transfer function $G^{CL}(z)$, i.e.

$$
z^4 + (h_3 - 3.98)z^3 + (h_2 + 5.94)z^2 + (h_1 - 3.94)z + h_0 + .98 = z^4 - 3.12z^3 \\
+ 3.61z^2 - 1.84z + .35
$$

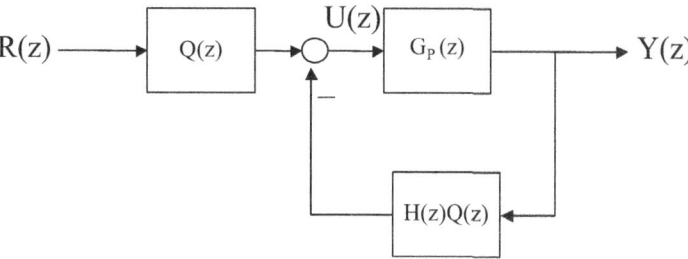

Fig. 3.6 System with zero-cancelling filter and feedback controller

This relationship leads to the following equations defining numerical values of parameters $h_0 - h_3$:

$h_3 - 3.98 = -3.12 \rightarrow h_3 = .86$
$h_2 + 5.94 = 3.61 \rightarrow h_2 = -2.33$
$h_1 - 3.94 = -1.84 \rightarrow h_1 = 2.10$
$h_0 + .98 = .35 \rightarrow h_0 = -.63$

Now that the controller $H(z)$ has been defined, let us modify the system block diagram as follows in Fig. 3.6.

It can easily be found that the overall transfer function of this system is

$$\frac{\dfrac{1}{.01(z^3 - 2.94z^2 + 2.88z - .94)} \times \dfrac{.01(z^3 - 2.94z^2 + 2.88z - .94)}{z^4 - 3.98z^3 + 5.94z^2 - 3.94z + 0.98}}{1 + \dfrac{.01(z^3 - 2.94z^2 + 2.88z - .94)}{z^4 - 3.98z^3 + 5.94z^2 - 3.94z + 0.98} \times \dfrac{1}{.01(z^3 - 2.94z^2 + 2.88z - .94)} \times (.86z^3 - 2.33z^2 + 2.10z - .63)}$$

$$= \frac{1}{(z^4 - 3.98z^3 + 5.94z^2 - 3.94z + 0.98) + (.86z^3 - 2.33z^2 + 2.10z - .63)}$$

$$= \frac{1}{z^4 - 3.12z^3 + 3.61z^2 - 1.84z + 0.35}$$

This indicates that the "last touch" of our design procedure is the modification of the filter in the reference channel. The modified filter should have the transfer function

$$F(z) = N_M(z)Q(z) = 6.76e - 05\left(z^3 + 8.94z^2 + 7.24z + .53\right)Q(z)$$

where $N_M(z)$ is the polynomial numerator of the model transfer function, then the overall system transfer function for the reference input will be exactly equal to $G_M(z)$.

Let us also define the closed-loop transfer function of the control system for the disturbance channel, understanding that the disturbance could be approximated by a staircase- type signal applied directly to the input of the controlled plant as shown below in Fig. 3.7. Then the closed loop transfer function of the disturbance channel is,

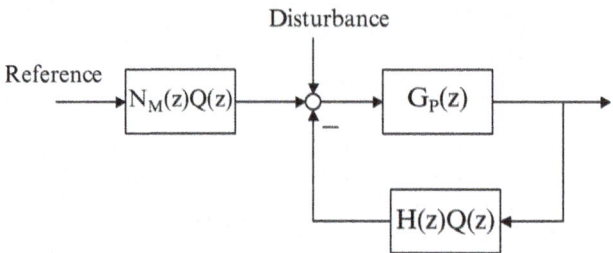

Fig. 3.7 System with input filter and feedback controller

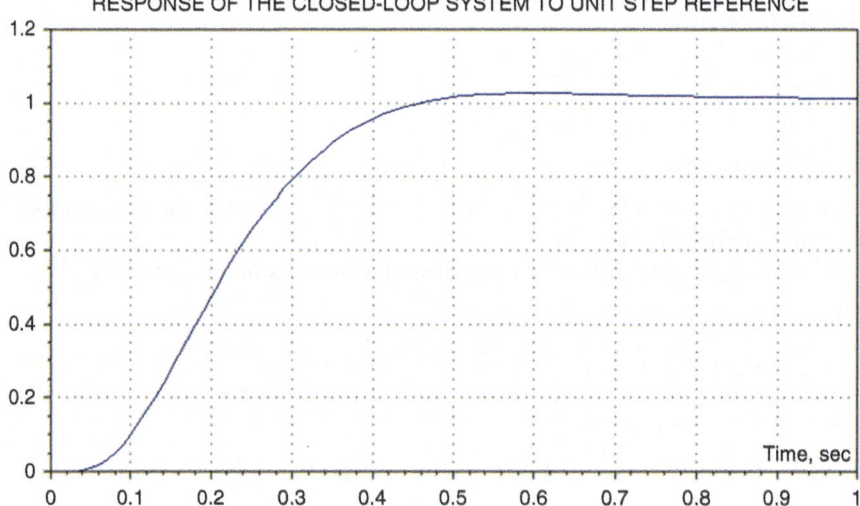

Fig. 3.8 Reference step response of closed-loop system

$$G^D(z) = \cfrac{\cfrac{.01\left(z^3 - 2.94z^2 + 2.88z - .94\right)}{z^4 - 3.98z^3 + 5.94z^2 - 3.94z + 0.98}}{1 + \cfrac{.01(z^3 - 2.94z^2 + 2.88z - .94)}{z^4 - 3.98z^3 + 5.94z^2 - 3.94z + 0.98} \times \cfrac{.86z^3 - 2.33z^2 + 2.10z - .63}{.01(z^3 - 2.94z^2 + 2.88z - .94)}}$$

$$= \frac{.01(z^3 - 2.94z^2 + 2.88z - .94)}{z^4 - 3.12z^3 + 3.61z^2 - 1.84z + .35}$$

The following Figs. 3.8, 3.9, 3.10, and 3.11 demonstrate the closed-loop characteristics of the design system

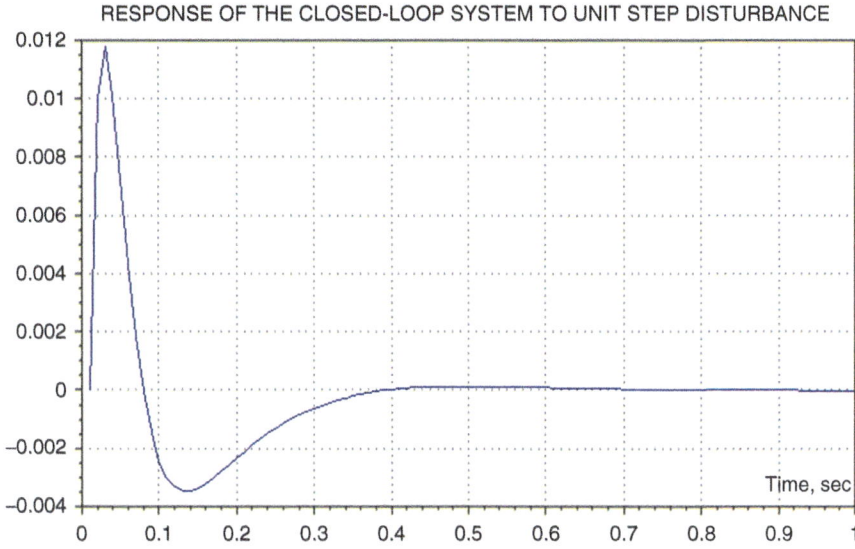

Fig. 3.9 Disturbance step response of closed-loop system

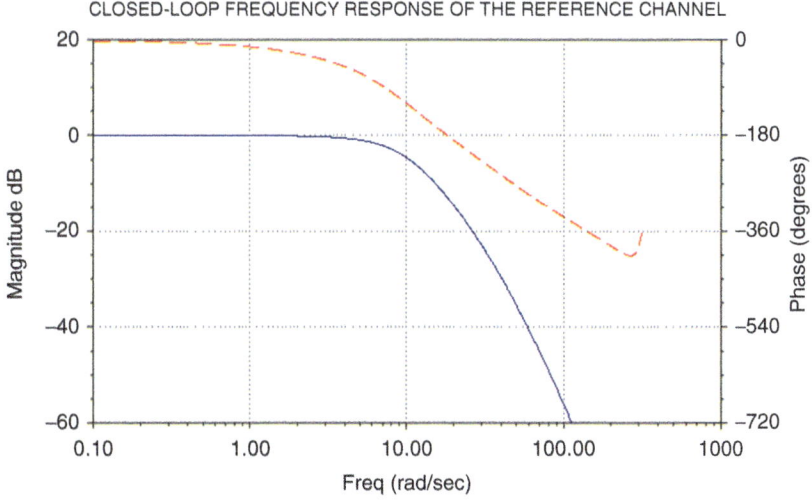

Fig. 3.10 Reference channel frequency response of closed-loop system

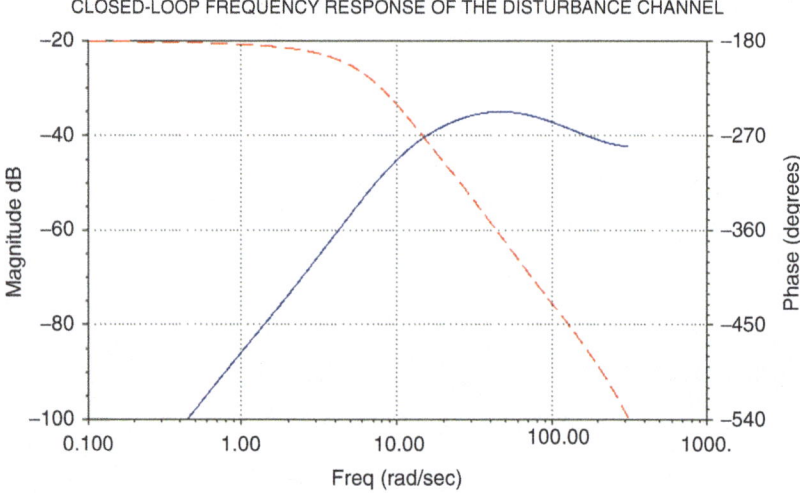

Fig. 3.11 Disturbance channel frequency response of closed-loop system

3.4 Control System with a State-Variable Feedback Controller

3.4.1 Discrete-Time State-Variable Description of Control Systems

In many instances the controlled process is described in continuous-time state-variable form,

$$\dot{X} = A_C X + B_C U, \text{ and } Y = CU$$

where $X = X(t)$ and $\dot{X} = \dot{X}(t)$ is the state vector $(n \times 1)$ comprising relevant physical variables (state variables) of the process and its first derivative, $Y = Y(t)$ is the vector $(m \times 1, m \le n)$ of controlled variables, $U = U(t)$ is the vector $(m \times 1)$ of controlled efforts, A_C is the fundamental matrix $(n \times n)$ of the system, B_C is the matrix $(n \times m)$ through which control efforts contribute to particular state equations, C is the matrix $(m \times n)$ that is used to designate the output variables, and subscript $_C$ is intended to remind the reader that we are dealing with a continuous-time system, and t is continuous time. It is expected state equations reflect laws of physics and state variables are real physical variables that can be continuously monitored through special sensors.

Development of a computer-based control system for a controlled plant represented by a state-variable model requires that its continuous-time description be converted into a discrete-time state-variable form,

$$X(k + 1) = AX(k) + BU(k)$$
$$Y(k) = CX(k), \ k = 0, 1, 2, ...$$

Again, the conversion must be performed for the required time step, T, and for the zero-order-hold application. Although the conversion should be accomplished using an engineering software tool, it is good to remember that

$$A = I + T \cdot A_C + \frac{T^2}{2} A_C^2 + \frac{T^3}{2 \cdot 3} A_C^3 + \frac{T^4}{2 \cdot 3 \cdot 4} A_C^4 + ...$$

$$B = \left[T \cdot I + \frac{T^2}{2} A_C + \frac{T^3}{2 \cdot 3} A_C^2 + \frac{T^4}{2 \cdot 3 \cdot 4} A_C^3 + ... \right] \cdot B_C$$

where I is the identity matrix, and matrix C is the same for the continuous- and discrete-time forms.

A state-variable controller implements the following control law:

$$U(k) = R(k) - FX(k), \ k = 1, 2, 3, ...$$

where U(k) is the control effort (m × 1) applied to the plant in the form of staircase-type signals, R(k) is the reference signal (m × 1), and F is a matrix (m × n) of parameters of the controller. Note that signal R(k) may be different from the set-point signal $R_0(k)$ actually representing the desired values of the output variables Y(k). As shown in the diagram below in Fig. 3.12, signal R(k) can be defined as $R(z) = W(z)R_0(z)$ where W(z) is a digital matrix-filter (m × m).

The matrix diagram above helps to realize that the control system has two different channels and two closed-loop transfer functions (transfer matrices), for the reference channel,

$$G_{CL}^R(z) = C(zI - A + BF)^{-1}BW(z)$$

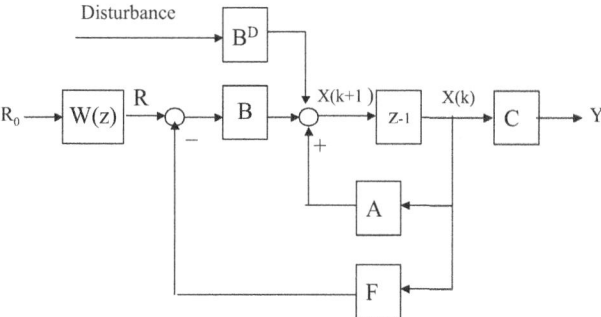

Fig. 3.12 State-variable definition of system with filter and feedback controller

and for the disturbance channel,

$$G_{CL}{}^D(z) = C(zI - A + BF)^{-1}B^D,$$

where B^D is the matrix through which the disturbance contributes to the state equations. Note that generally speaking, matrix B^D is different from matrix B.

Indeed, combining the control law with the state-variable description of the controlled plant, results in the state-variable description of the closed-loop system,

$$X(k + 1) = AX(k) + BR(k) - BFX(k) \text{ and } Y(k) = CX(k), \ k = 0, 1, 2, \dots.$$

Taking the Z-transform of these equations under zero initial conditions results in

$$zX(z) = AX(z) + BR(z) - BFX(z) \text{ and } Y(z) = CX(z)$$
$$zX(z) = AX(z) + BR(z) - BFX(z) \text{ or } (zI - A + BF)X(z) = BR(z), \text{ and}$$
$$X(z) = (zI - A + BF)^{-1}BR(z)$$

Multiplying left and right-hand sides of the last equation by matrix C results in the relationship between the reference and the output vector,

$$CX(z) = C(zI - A + BF)^{-1}BR(z) = Y(z)$$

Recall that $R(z) = W(z)R_0(z)$, then

$$Y(z) = C(zI - A + BF)^{-1}BW(z)R_0(z)$$

and consequently, $C(zI - A + BF)^{-1}BW(z)$ should be interpreted as the transfer function (transfer matrix) of the reference channel.

In the case of the disturbance input that may contribute to the state equations through matrix $B^D \neq B$, the situation is as follows:

$$X(k + 1) = AX(k) + BU(k) + B^D D(k)$$
$$U(k) = R(k) - FX(k)$$
$$Y(k) = CX(k), \qquad k = 0, 1, 2, \dots.$$

$$X(k + 1) = AX(k) + BR(k) - BFX(k) + B^D D(k)$$
$$Y(k) = CX(k), \qquad k = 0, 1, 2, \dots.$$

$$zX(z) = AX(z) + BR(z) - BFX(z) + B^D D(z), \ Y(z) = CX(z)$$
$$(zI - A + BF)X(z) = BR(z) + B^D D(z)$$
$$X(z) = (zI - A + BF)^{-1}BR(z) + (zI - A + BF)^{-1}B^D D(z)$$

Multiply the left-hand side and right-hand side of the equation by C and for the disturbance input assume $R(z) = 0$, then

$$Y(z) = C(zI - A + BF)^{-1}B^D D(z)$$

and finally the transfer function (transfer matrix) of the disturbance channel is

$$C(zI - A + BF)^{-1}B^D$$

$$X(k + 1) = A_M X(k) + B_M U(k), \quad Y(k) = C_M X(k), \quad k = 0, 1, 2, ...$$

In many instances the state-variable description of the control system in its arbitrary form (most likely consistent with the laws of physics or results of regression analysis)

$$X(k + 1) = AX(k) + BU(k) + B^D D(k)$$
$$U(k) = R(k) - FX(k)$$
$$Y(k) = CX(k), \qquad k = 0, 1, 2, ...$$

should be converted into the canonical controllable form (CCF),

$$V(k + 1) = A_{CCF}V(k) + B_{CCF}U(k) + B^D{}_{CCF}D(k)$$
$$U(k) = R(k) - F_{CCF}V(k)$$
$$Y(k) = C_{CCF}V(k), \qquad k = 0, 1, 2, ...$$

where $A_{CCF} = PAP^{-1}$, $B_{CCF} = PB$, $B^D{}_{CCF} = PB^D$, $F_{CCF} = FP^{-1}$, $C_{CCF} = CP^{-1}$, $V(k) = PX(k)$, and P is a $n \times n$ matrix, providing the "key" to converting an arbitrary state-variable form to the CCF. As per undergraduate controls,

$$P = \begin{bmatrix} Q \\ QA \\ QA^2 \\ \cdots \\ QA^{n-1} \end{bmatrix} \text{ where } Q = \begin{bmatrix} 0 & 0 & \cdots & 0 & 1 \end{bmatrix} \cdot \begin{bmatrix} B & AB & A^2B & \cdots & A^{n-1}B \end{bmatrix}^{-1}$$

Let us verify the relationship between V(k) and X(k) state vectors (assume D(k) = 0):

$$V(k + 1) = PAP^{-1}V(k) + PBU(k) \text{ or } P^{-1}V(k + 1) = AP^{-1}V(k) + BU(k)$$

Indeed, if $P^{-1}V(k) = X(k)$ the last equation turns into the original state equation in the arbitrary form, $X(k + 1) = AX(k) + BU(k)$

3.4.2 Design of a Discrete-Time State-Variable Controller

Assume that the design specifications are given by the settling time, T_{SET}, overshoot of the step response, P%, and the steady-state error caused by a unit step disturbance. Consider a design procedure resulting in a control system with a state-variable controller consistent with the above specifications. The procedure is intended for the situation when the transfer function of the controlled plant,

$$G_P(z) = C(zI - A)^{-1}B$$

does not have zeros outside the unit circle.

First, it is recommended to define an s-domain model transfer function $G_M(s)$ in full compliance with the settling time and the overshoot of the step response that can be achieved by the choice of dominant and non-dominant poles. The order of this transfer function must be the same as the order of the controlled plant, and it shall not have zeros. The constant gain of this transfer function must be chosen such that $G_M(0) = 1$.

The next step is to convert model transfer function into the discrete-time domain using the ZOH option and the appropriate time step, T, that will result in transfer function $G_M(z)$:

$$G_M(s) \underset{ZOH}{\overset{T}{\rightarrow}} G_M(z)$$

Finally, $G_M(z)$ should be subjected to direct decomposition that will yield the state-variable equivalent of $G_M(z)$ in the canonical controllable form (CCF),

$$X(k+1) = A_M X(k) + B_M U(k), \ \ Y(k) = C_M X(k), \ \ k = 0, 1, 2, ...$$

Convert the state-variable description of the control system

$$\begin{aligned}X(k+1) &= AX(k) + BU(k) + B^D D(k) \\ U(k) &= R(k) - FX(k) \\ Y(k) &= CX(k), \qquad k = 0, 1, 2, ...\end{aligned}$$

into the CCF form,

$$\begin{aligned}V(k+1) &= A_{CCF} V(k) + B_{CCF} U(k) + B^D{}_{CCF} D(k) \\ U(k) &= R(k) - F_{CCF} V(k) \\ Y(k) &= C_{CCF} V(k), \qquad k = 0, 1, 2, ...\end{aligned}$$

This conversion would drastically simplify the controller design problem.

It can be seen that the state equation of the closed-loop system is

$$V(k+1) = (A_{CCF} - B_{CCF}F_{CCF})V(k) + B_{CCF}R(k) + B^D{}_{CCF}D(k)$$

It is our goal to assure that the fundamental matrix of the closed-loop system in CCF, $A_{CCF}-B_{CCF}F_{CCF}$, be equal to the fundamental matrix of the model system, A_M, i.e.

$$A_{CCF} - B_{CCF}F_{CCF} = A_M$$

Therefore, $B_{CCF}F_{CCF} = A_{CCF} - A_M$. Generally speaking, this equation is not "user-friendly", especially when B_{CCF} is not a square matrix, however, in the case of a single-input-single-output system and a CCF format of all relevant matrices, this equation looks like,

$$\begin{bmatrix} 0 \\ 0 \\ \cdots \\ 0 \\ 1 \end{bmatrix} \begin{bmatrix} f_1 & f_2 & \cdots & f_{n-1} & f_n \end{bmatrix} = \begin{bmatrix} 0 & 1 & 0 & \cdots & 0 \\ 0 & 0 & 1 & \cdots & 0 \\ \cdots & \cdots & \cdots & \cdots & \cdots \\ 0 & 0 & 0 & \cdots & 1 \\ a_1{}^{CCF} & a_2{}^{CCF} & a_3{}^{CCF} & \cdots & a_n{}^{CCF} \end{bmatrix}$$

$$- \begin{bmatrix} 0 & 1 & 0 & \cdots & 0 \\ 0 & 0 & 1 & \cdots & 0 \\ \cdots & \cdots & \cdots & \cdots & \cdots \\ 0 & 0 & 0 & \cdots & 1 \\ a_1{}^M & a_2{}^M & a_3{}^M & \cdots & a_n{}^M \end{bmatrix}$$

and consequently,

$$F_{CCF} = B_{CCF}{}^T(A_{CCF} - A_M) = \begin{bmatrix} 0 & 0 & \cdots & 0 & 1 \end{bmatrix}(A_{CCF} - A_M)$$

The designed controller is intended for the state vector of the CCF, $V(k)$, to provide the state-variable feedback. It is now time to obtain the controller consistent with the "real" state vector, $X(k)$, by post-multiplying matrix F_{CCF} by matrix P:

$$F = F_{CCF}P$$

It can be found that closed-loop transfer function

$$C(zI - A + BF)^{-1}B$$

has the same denominator (characteristic polynomial) as the model transfer function, $G_M(z)$. However, its numerator, $N(z) = N_P(z)$, is equal to the numerator of the transfer function of the controlled plant, $G_P(z) = C(zI - A)^{-1}B$ and is different from the numerator of the model transfer function $N_M(z)$. This reality will adversely affect the overshoot of the step response of the designed system, and can be corrected by the introduction of a digital filter in the reference channel of the designed system,

$$W(z) = \frac{N_M(z)}{N_P(z)}$$

Finally, the steady-state error caused by unit step disturbance applied to the designed system can be evaluated. Recall that the error caused by disturbance is the system response to this disturbance. The z-transform of the error caused by unit step disturbance, $E(z)$, can be defined as the z-transform of the unit step signal multiplied by the transfer function of the disturbance channel:

$$E(z) = C(zI - A + BF)^{-1}B^D\frac{z}{z-1}$$

The steady-state error, E_{SS}, is defined by the application of the final value theorem of z-transform to the above expression:

$$E_{SS} = \lim_{k\to\infty} E(k) = \lim_{z\to1} \frac{z-1}{z}E(z) = \lim_{z\to1} \frac{z-1}{z}C(zI - A + BF)^{-1}B^D\frac{z}{z-1}$$
$$= C(I - A + BF)^{-1}$$

According to this result, the value of the steady-state error, E_{SS}, can be manipulated by the choice of non-dominant poles of the model transfer function $G_M(s)$ in the beginning of the design procedure.

Example 3.5 The following are the state and output equations of the controlled plant obtained on the basis of regression models, note explicitly defined matrices A, B, and C:

$$X(k+1) = \begin{bmatrix} -0.440 & 4.095 & 4.716 & 4.072 \\ -0.333 & 3.090 & 2.730 & 1.468 \\ -0.331 & -2.612 & -1.164 & 0.408 \\ -0.440 & 2.716 & 2.851 & 2.504 \end{bmatrix} X(k) + \begin{bmatrix} 2 \\ 0 \\ 1 \\ 2 \end{bmatrix} U(k) \text{ and } Y(k)$$
$$= [-1 \ -0.3 \ 1 \ 3] \times 1e - 5$$

The design specifications call for 0.67 s settling time, overshoot of the step response of 10 %, and the steady-state error for unit step disturbance $E_{SS} \leq 0.15$ (abs. units). The time step of the digital controller is 0.005 s.

The following model transfer function is defined in the s-domain in compliance with the settling time and overshoot requirements:

$$G_M(s) = \frac{200000}{(s^2 + 12s + 100)(s + 40)(s + 50)}$$

It has been converted into the z-domain for the time step of 0.005 s using the zero-order-hold option:

$$G_M(z) = \frac{4.707e - 06(z^3 + 9.947z^2 + 8.982z + 0.7364)}{(z^2 - 1.939z + 0.9418)(z - 0.8187)(z - 0.7788)}$$

The direct decomposition of the above transfer function results in the following fundamental matrix:

$$A_M = \begin{bmatrix} 0 & 1 & 0 & 0 \\ 0 & 0 & 1 & 0 \\ 0 & 0 & 0 & 1 \\ -0.600 & 2.741 & -4.678 & 3.537 \end{bmatrix}$$

The following matrix filter facilitating the conversion of the state-variable description of the controlled plant to CCF was obtained:

$$P = \begin{bmatrix} 0.222 & -0.302 & -0.090 & -0.177 \\ 0.111 & -0.273 & -0.180 & -0.021 \\ 0.111 & 0.025 & -0.072 & -0.075 \\ -2e - 04 & 0.514 & 0.460 & 0.270 \end{bmatrix}$$

The following matrix is the result of converting the fundamental matrix of the controlled plant to CCF, i.e. $A_{CCF} = PAP^{-1}$:

$$A_{CCF} = \begin{bmatrix} 0 & 1 & 0 & 0 \\ 0 & 0 & 1 & 0 \\ 0 & 0 & 0 & 1 \\ -0.991 & 3.973 & -5.974 & 3.990 \end{bmatrix}$$

Since $A_{CCF} - B_{CCF}F_{CCF} = A_M$, due to specific configuration of matrix B_{CCF} the matrix of the controller F_{CCF} can be found as the "last row of the difference $A_{CCF} - A_M$":

$$F_{CCF} = \begin{bmatrix} 0 & 0 & 0 & 1 \end{bmatrix}(A_{CCF} - A_M) = \begin{bmatrix} -0.390 & 1.232 & -1.296 & 0.453 \end{bmatrix}$$

Finally, matrix F of the state-variable controller consistent with the state vector $X(k)$ is defined as follows:

$$F = F_{CCF}P = \begin{bmatrix} -0.094 & -0.017 & 0.115 & 0.263 \end{bmatrix}$$

Define the closed-loop transfer function $G_{CL}(z) = C(zI - A + BF)^{-1}B$:

$$G_{CL}(z) = \frac{5e - 5(z - 0.6002)(z^2 - 2.1z + 1.802)}{(z - 0.7788)(z - 0.8187)(z^2 - 1.939z + 0.9418)}$$

One can realize that it has the denominator of the model transfer function, as required, but its numerator is quite different from the numerator of the model transfer function. Note that the computation of the closed-loop transfer function is quite a formidable task; it is much easier to define this numerator from the matrix C_{CCF} of the plant:

$$C_{CCF} = [-5.407 \quad 15.31 \quad -13.50 \quad 5.000]1e - 5$$

$$N_P(z) = 1e - 5(5z^3 - 13.50z^2 + 15.31z - 5.407)$$

$$W(z) = \frac{4.707e - 06(z^3 + 9.947z^2 + 8.982z + 0.7364)}{1e - 5(5z^3 - 13.50z^2 + 15.31z - 5.407)}$$

or

$$W(z) = \frac{0.4707(z^3 + 9.947z^2 + 8.982z + 0.7364)}{5z^3 - 13.50z^2 + 15.31z - 5.407}$$

The simulation indicates that the reference channel of the designed system is compliant with the design specifications (see Fig. 3.13).

Application of the final value theorem results in the following value of the steady-state error for unit step disturbance (it is assumed that $B^D = B$):

$$E_{SS} = C(I - A + BF)^{-1}B = 0.144 \quad (units)$$

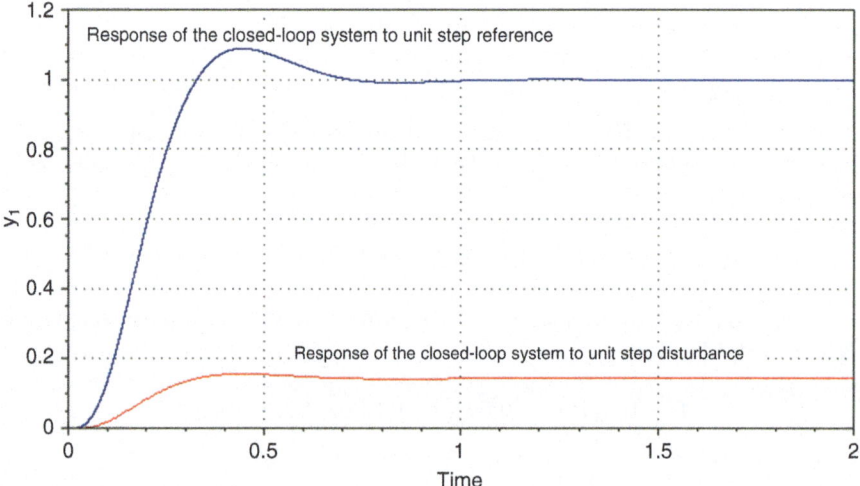

Fig. 3.13 System response to step reference and step disturbance

Exercise 3.1

Problem 1 The controlled plant is defined by its transfer function,

$$G_P(s) = \frac{s^2 + 6s + 10}{s^3 + 2s^2 + 9s + 8}$$

Utilize a simulator to obtain the response of the controlled plant to a combination of random noise and a sinusoidal signal. Record 500 successive measurements of the resultant input/output data discretized with the time step of 0.05 s. Obtain the z-domain transfer function of the controlled plant, $G_P(z)$, using the LSM. Verify your result using a software tool performing the conversion $G_P(s) \xrightarrow[\text{ZOH}]{0.05 \ \text{sec}} G_P(z)$.

Problem 2 Utilize the transfer function, $G_P(z)$, obtained in the above problem to design a discrete-time output-feedback control system compliant with the following specifications: settling time $T_{SET} = 2$ s, overshoot of a step response $P\% = 5\,\%$, and the steady-state error for unit-step disturbance $E_{SS} \leq 0.2$ (units). Show all stages of your design. Investigate the design by computer simulation.

Problem 3 The controlled plant is defined by the following continuous-time equations:

$$\dot{x}_1 = -10.4x_1 + 10.3x_2 + 8.8x_3 - 3u$$
$$\dot{x}_2 = 0.6x_1 - 0.65x_2 - 0.2x_3 + u$$
$$\dot{x}_3 = -11.9x_1 + 11.7x_2 + 9.6x_3 - 4u$$
$$y = -7.3x_1 + 7.6x_2 + 6.8x_3$$

Obtain the discrete-time equivalent of these equations for the time step of 0.005 s and zero-order-hold application. Utilize 6 terms of the series approximation.

Problem 4 Utilize the solution of the previous problem to design a discrete-time state-variable feedback control system compliant with the following specifications: settling time $T_{SET} = 2$ s, overshoot of a step response $P\% = 5\%$, and the steady-state error for unit-step disturbance $E_{SS} \leq 0.2$ (units). Show all stages of your design.

3.5   Control System with a State-Variable Feedback and a State Observer

3.5.1   Discrete-Time State-Variable Control Systems with State Observers

Discrete-time state-variable control has a number of advantages over the output-feedback control scheme. Due to these advantages, it can be utilized even when state-variables of the controlled plant are not accessible. This is accomplished by the use of a state observer, a software-based dynamic system capable of computing the estimated state vector using the input and output data of the controlled plant:

$$Z(i+1) = MZ(i) + BU(i) + KY(i), \quad i = 1, 2, 3, \ldots$$

where M is a fundamental matrix of the state observer and I is the discrete-time index. It can be seen that the control effort, U(i), and the plant output, Y(i), are playing the role of the "forcing functions" for the observer. Under some conditions, the state vector of the observer, Z(i), having the same dimensions as the unavailable state vector of the controlled plant, X(i), converges to X(i), i.e. the state observation error, $E(i) = X(i) - Z(i)$, is such that $\lim_{i \to \infty} E(i) = 0$.

Consider the full description of a controlled plant with a state observer and a state-variable controller:

$$\begin{aligned} X(i+1) &= AX(i) + BU(i) \\ Z(i+1) &= MZ(i) + BU(i) + KY(i) \\ U(i) &= R(i) - FZ(i) \\ Y(i) &= CX(i), \qquad i = 0, 1, 2, \ldots. \end{aligned}$$

Subtracting the second state equation from the first yields:

$$\begin{aligned} X(i+1) - Z(i+1) &= AX(i) + BU(i) - MZ(i) - BU(i) - KY(i) \quad \text{or} \\ X(i+1) - Z(i+1) &= AX(i) + BU(i) - MZ(i) - BU(i) - KCX(i) \quad \text{or} \\ E(i+1) &= (A - KC)X(i) - MZ(i) \end{aligned}$$

Define matrix M as A-KC, then

$$\begin{aligned} E(i+1) &= (A - KC)X(i) - MZ(i) = (A - KC)X(i) - (A - KC)Z(i) \quad \text{or} \\ E(i+1) &= (A - KC)E(i) \end{aligned}$$

The resultant equation describes the error conversion (or the state estimation) process. It is known that to assure that $\lim_{i \to \infty} E(i) = 0$, one has to properly assign the eigenvalues of the fundamental matrix of the state estimation process, A-KC. Since it is a discrete-time process the eigenvalues must be chosen inside the unit circle in the complex plane, i.e.

$$|\lambda_k| < 1 \quad \text{for all } k = 1, 2, \ldots, n$$

One can realize that this can be accomplished by the choice of matrix K.

Now transform the state equation of the closed-loop system as follows:

$$X(i+1) = AX(i) + BR(i) - BFZ(i)$$

Since $E(i) = X(i) - Z(i)$, $Z(i) = X(i) - E(i)$,

$$X(i + 1) = AX(i) + BR(i) - BFX(i) + BFE(i) \text{ or } X(i + 1)$$
$$= (A - BF)X(i) + BFE(i) + BR(i)$$

Combining the last equation with the equation of the estimation process, one can obtain the complete state-variable description of the control system with the state-variable controller and the state observer that would describe both the control and the state estimation processes:

$$X(i + 1) = (A - BF)X(i) + BFE(i) + BR(i)$$
$$E(i + 1) = (A - KC)E(i)$$

or in a block-matrix form

$$\begin{bmatrix} X(i + 1) \\ E(i + 1) \end{bmatrix} = \begin{bmatrix} A - BF & BF \\ \varnothing & A - KC \end{bmatrix} \begin{bmatrix} X(i) \\ E(i) \end{bmatrix} + \begin{bmatrix} B \\ \varnothing \end{bmatrix} R(i)$$

This result demonstrates the separation principle: the estimation process is completely independent from the control process. However, it can be seen that the estimation error may affect the control process during the transient regime but not in steady-state, since $\lim_{i \to \infty} E(i) = 0$.

It shall not be forgotten that a special filter $W(z)$ in the reference channel and a filter $H(z)$ in the input of the controlled plant may be required to achieve the full compliance of the designed system with the model transfer function $G_M(z)$ and simplify the design procedure. The final configuration of a discrete-time control system with a state-variable controller and a state observer is shown below in Fig. 3.14.

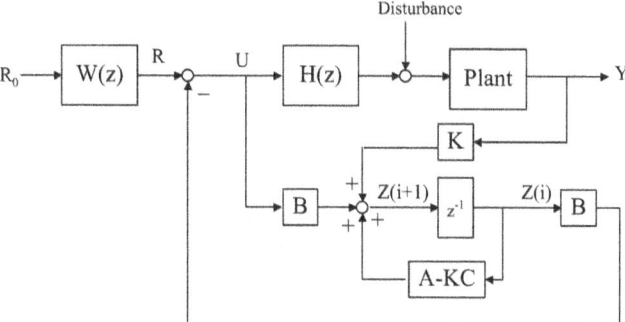

Fig. 3.14 State-variable system with state observer

3.5.2   Design and Analysis

Contribution of a state observer in the control process can be seen from the following, previously obtained equations,

$$X(i + 1) = (A - BF)X(i) + BFE(i) + BR(i)$$
$$E(i + 1) = (A - KC)E(i)$$

One can conclude that state estimation error shall not interfere with the control process. This can be achieved by making its convergence time much shorter that the settling time of the control system and eigenvalues of matrix A-KC should be chosen appropriately.

Consider the computational task of defining the matrix

$$K = \begin{bmatrix} K_1 \\ K_2 \\ \cdots \\ K_n \end{bmatrix}$$

First note that matrix A, matrix B and matrix C represent the CCF of the controlled plant's mathematical description. Since the state vector of the plant is inaccessible and is to be estimated, it makes a lot of sense to estimate it for the CCF—the most convenient state-variable form. Matrix K must be chosen such that eigenvalues of the matrix A-KC be equal to the desired eigenvalues $\lambda_1, \lambda_2, \ldots, \lambda_n$ that would assure the necessary convergence rate of state estimation errors. This requirement leads to the equation

$$Det(zI - A + KC) = (z - \lambda_1) \cdot (z - \lambda_2) \cdot \ldots \cdot (z - \lambda_n) = z^n + p_{n-1}z^{n-1}$$
$$+ p_{n-2}z^{n-2} + \ldots + p_1 z + p_0$$

The left-hand side of this equation, in more detail, looks like this:

$$Det\left(\begin{bmatrix} z & 0 & 0 & \cdots & 0 \\ 0 & z & 0 & \cdots & 0 \\ 0 & 0 & z & \cdots & 0 \\ \cdots & \cdots & \cdots & \cdots & \cdots \\ 0 & 0 & 0 & \cdots & z \end{bmatrix} - \begin{bmatrix} 0 & 1 & 0 & \cdots & 0 \\ 0 & 0 & 1 & \cdots & 0 \\ \cdots & \cdots & \cdots & \cdots & \cdots \\ 0 & 0 & 0 & \cdots & 1 \\ a_{n1} & a_{n2} & a_{n3} & \cdots & a_{nn} \end{bmatrix} + \begin{bmatrix} K_1 \\ K_2 \\ K_3 \\ \cdots \\ K_n \end{bmatrix} \begin{bmatrix} c_1 & c_2 & c_3 & \cdots & c_n \end{bmatrix} \right)$$

and constitutes a n-order polynomial with respect to z. While matrix A and matrix C are known, coefficients of z in the powers of $n-1, n-2, \ldots, 1, 0$ in this polynomial are functions of $K_1, K_2, \ldots, K_n$. Technically, these coefficients can be equated to the corresponding coefficients of the polynomial $z^n + p_{n-1}z^{n-1} + p_{n-2}z^{n-2} + \ldots + p_1 z + p_0$ resulting in the system of n equations with n unknowns, $K_1, K_2, \ldots, K_n$. This concept has one major flaw: it can be seen that these equations will be non-linear with respect to unknowns and their solution will present a formidable task.

We propose the following approach to overcome this difficulty. Introduce a digital filter in the input of the controlled plant as shown in the block diagram above,

$$H(z) = \frac{1}{N_P(z)}$$

where $N_P(z)$ is the numerator of the controlled plant transfer function. Now the CCF description of the controlled plant & filter will have the same matrix A and matrix B as the plant alone, but its matrix C will be replaced by

$$\overline{C} = \begin{bmatrix} 1 & 0 & \cdots & 0 & 0 \end{bmatrix}$$

Now the polynomial $\text{Det}\left(zI - A + K\overline{C}\right)$ looks as follows,

$$\text{Det}\left(\begin{bmatrix} z + K_1 & -1 & 0 & \cdots & 0 \\ K_2 & z & -1 & \cdots & 0 \\ K_3 & 0 & z & \cdots & 0 \\ \cdots & \cdots & \cdots & \cdots & \cdots \\ -a_{n1} + K_n & -a_{n2} & -a_{n3} & \cdots & z - a_{nn} \end{bmatrix}\right) = Q(z, K)$$

and coefficients of the resultant polynomial

$$Q(z, K) = z^n + \hat{p}_{n-1}(K)z^{n-1} + \hat{p}_{n-2}(K)z^{n-2} + \ldots + \hat{p}_1(K)z + \hat{p}_0(K)$$

are linear with respect to unknown elements of matrix K.

When the state observer matrix K has been established, the design of the state-variable controller matrix F is very consistent with the previous section and Example 3.5.

Finally, consider matrix equations describing the entire control system,

$$\begin{bmatrix} X(i+1) \\ E(i+1) \end{bmatrix} = A_{CL} \begin{bmatrix} X(i) \\ E(i) \end{bmatrix} + B_{CL} R(i)$$

$$Y(i) = C_{CL} \begin{bmatrix} X(i) \\ E(i) \end{bmatrix},$$

where

$$A_{CL} = \begin{bmatrix} A - BF & BF \\ \varnothing & A - K\overline{C} \end{bmatrix}, \quad B_{CL} = \begin{bmatrix} B \\ \varnothing \end{bmatrix}, \quad \text{and} \quad C_{CL} = \begin{bmatrix} \overline{C} & \varnothing \end{bmatrix}$$

Note that in these equations matrix C is replaced by $\overline{C} = \begin{bmatrix} 1 & 0 & \cdots & 0 & 0 \end{bmatrix}$ to account for the cancelled zeros of the plant.

In order to finalize the design and, later, to subject the resultant system to steady-state error analysis, obtain the transfer functions of this system for the reference channel:

$$G^R_{CL}(z) = C_{CL}(zI - A_{CL})^{-1}B_{CL}$$

In the situation when the system is properly designed, the first transfer function

$$G^R_{CL}(z) = \frac{1}{D_M(z)}$$

where $D_M(z)$ is the denominator of the model transfer function. Therefore, in order to comply with design specifications it is required to define the input signal $R(z)$ as

$$R(z) = R_0(z)W(z) = R_0(z)N_M(z)$$

where $R_0(z)$ is the set point signal (or control input), $W(z)$ is a filter in the reference channel, and $N_M(z)$ is the numerator of the model transfer function.

To perform the analysis of the steady-state error caused by disturbance, typically, a unit-step disturbance, a transfer function of the disturbance channel, $G^D_{CL}(z)$, should be established. Since the disturbance $D(z) = \frac{z}{z-1}$, according to the z-domain final value theorem, the steady-state error for unit step disturbance is defined as

$$E_{SS} = \lim_{k \to \infty} E(k) = \lim_{z \to 1} \frac{z-1}{z} E(z) = \lim_{z \to 1} \frac{z-1}{z} G^D_{CL}(z) \frac{z}{z-1} = G^D_{CL}(1)$$

or

$$E_{SS} = C^E(I - A^E)^{-1}B^E$$

where A^E, B^E and C^E are matrices of the CCF obtained by the direct decomposition of the transfer function $G^D_{CL}(z)$. However, derivation of this transfer function is not a straightforward task. It is "safe" therefore to consider the following interpretation of the block diagram of a control system with state-variable feedback and a state observer seen in Fig. 3.15:

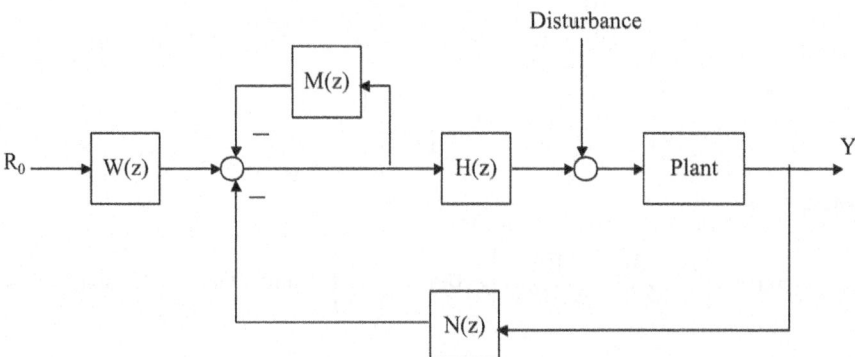

Fig. 3.15 Control system with state-variable feedback and state observer

where

$$M(z) = F(zI - A_P + K\overline{C})^{-1}B_P \text{ and } N(z) = F(zI - A_P + K\overline{C})^{-1}K$$

Then the transfer function of the disturbance channel can be defined as

$$G^D_{CL}(z) = \frac{Y(z)}{D(z)} = \frac{G_P(z)}{1 + G_P(z)H(z)N(z)\frac{1}{1+M(z)}} = \frac{G_P(z)[1 + M(z)]}{1 + M(z) + G_P(z)H(z)N(z)}$$

thus providing the basis for the steady-state error analysis.

Example 3.6 Given transfer function of a controlled plant, $G_P(s) = \dfrac{3s^2 + 5s + 10}{s^3 + 2s^2 + 3s + 5}$. It is required to design a discrete-time control system with a state-variable controller operating with the clock frequency of 0.005 s to achieve the settling time of 5 s, overshoot of the step response of 10% and the steady-state error for a unit-step disturbance of 0.025 abs units. Since the state variables of the controlled plant are not accessible, a state observer must be designed as well.

First, let us convert G(s) into the z-domain using the zero-order-hold option and the time step T = 0.005 s:

$$G_P(z) = \frac{0.01499(z^2 - 1.992z + 0.9917)}{(z^2 - 1.999z + 0.9992)(z - 0.9908)}$$

and obtain matrices of the CCF of this expression:

$$A_P = \begin{bmatrix} 0 & 1 & 0 \\ 0 & 0 & 1 \\ 0.990 & -2.980 & 2.990 \end{bmatrix}, \ B_P = \begin{bmatrix} 0 \\ 0 \\ 1 \end{bmatrix}, \ C_P = [0.015 \ -0.030 \ 0.015]$$

Introduce filter $H(z) = \frac{1}{0.01499(z^2 - 1.992z + 0.9917)}$ in the input of the controlled plant, the CCF description of the *filter & plant* is as follows:

$$A_P = \begin{bmatrix} 0 & 1 & 0 \\ 0 & 0 & 1 \\ 0.990 & -2.980 & 2.990 \end{bmatrix}, \ B_P = \begin{bmatrix} 0 \\ 0 \\ 1 \end{bmatrix}, \ C_P = [1 \ 0 \ 0]$$

Introduce an s-domain model transfer function consistent with the required closed-loop dynamics of the reference channel:

$$G_M(s) = \frac{7.076}{(s^2 + 1.596s + 1.769)(s + 4)}$$

Conversion of $G_M(s)$ into the z-domain using the zero-order-hold option and the time step T = 0.005 s and consequent direct decomposition of this transfer function yield:

$$G_M(z) = \frac{1.464e - 07(z^2 + 3.972z + 0.9861)}{(z^2 - 1.992z + 0.9921)(z - 0.9802)}$$

$$A_M = \begin{bmatrix} 0 & 1 & 0 \\ 0 & 0 & 1 \\ 0.972 & -2.945 & 2.972 \end{bmatrix}, \; B_M = \begin{bmatrix} 0 \\ 0 \\ 1 \end{bmatrix}, \; C_M = \begin{bmatrix} 1 & 6 & 1 \end{bmatrix} \times 1e - 7$$

Introduce an s-domain transfer function representing the desired dynamics of the on process (note that its transient terms are approximately 4 times faster then the non-dominant term of the control process):

$$G_E(s) = \frac{1}{(s + 15)(s + 16)(s + 18)}$$

Conversion of $G_E(s)$ into the z-domain using the zero-order-hold option and the time step $T = 0.005$ s yields:

$$G_E(z) = \frac{1.96e - 08(z^2 + 3.763z + 0.8847)}{(z - 0.9277)(z - 0.9231)(z - 0.9139)}$$

While the numerator of this transfer function is meaningless, the denominator exhibits three desired eigenvalues of the fundamental matrix of the state observer, $\lambda_1 = 0.9277$, $\lambda_2 = 0.9231$, and $\lambda_3 = 0.9139$. As a matter of fact,

$$\text{Det}\left(zI - \begin{bmatrix} 0 & 1 & 0 \\ 0 & 0 & 1 \\ 0.990 & -2.980 & 2.990 \end{bmatrix} + \begin{bmatrix} K_1 \\ K_2 \\ K_3 \end{bmatrix} \begin{bmatrix} 1 & 0 & 0 \end{bmatrix} \right)$$
$$= (z - 0.9277)(z - 0.9231)(z - 0.9139)$$

or

$$\text{Det} \begin{bmatrix} z + K_1 & -1 & 0 \\ K_2 & z & -1 \\ K_3 - 0.990 & 2.980 & z - 2.990 \end{bmatrix} = z^3 - 2.765z^2 + 2.548\,z - 0.783$$

Transform the determinant in the left-hand of the equation as follows:

$$z(z + K_1)(z - 2.990) + (K_3 - 0.990) + 2.980(z + K_1) + K_2(z - 2.990) =$$
$$z^3 + z^2(K_1 - 2.990) - 2.990zK_1 + K_3 - 0.990 + 2.980z + 2.980K_1 + zK_2$$
$$- 2.990K_2 =$$

$$z^3 + z^2(K_1 - 2.990) + z(-2.990K_1 + 2.980 + K_2)$$
$$+ (K_3 - 0.990 + 2.980K_1 - 2.990K_2)$$

Now it can be stated that

$$K_1 - 2.990 = -2.765$$
$$-2.990K_1 + 2.980 + K_2 = 2.548$$
$$K_3 - 0.990 + 2.980K_1 - 2.990K_2 = -0.783$$

Note that these equations are linear with respect to K_1, K_2 and K_3 and their solution is very straightforward:

$$K = \begin{bmatrix} K_1 \\ K_2 \\ K_3 \end{bmatrix} = \begin{bmatrix} .225 \\ .241 \\ .258 \end{bmatrix}$$

The following recursive formula, obtained by the author for his students, facilitates easy computation of the matrix K for any size of the problem. For example, for $n = 3$:

$$K_1 = -A_E(3,3) + A_P(3,3)$$
$$K_2 = -A_E(3,2) + A_P(3,2) + A_P(3,3)*K_1$$
$$K_3 = -A_E(3,1) + A_P(3,1) + A_P(3,2)*K_1 + A_P(3,3)*K_2$$

For $n = 4$:

$$K_1 = -A_E(4,4) + A_P(4,4)$$
$$K_2 = -A_E(4,3) + A_P(4,3) + A_P(4,4)*K_1$$
$$K_3 = -A_E(4,2) + A_P(4,2) + A_P(4,3)*K_1 + A_P(4,4)*K_2$$
$$K_4 = -A_E(4,1) + A_P(4,1) + A_P(4,2)*K_1 + A_P(4,3)*K_2 + A_P(4,4)*K_1$$

and etc.

Generally a software tool, such as program CC, is used for the "automated" solution for the observer matrix. The utilization of a software tool does not require canceling the zeros of the controlled plant and should be recommended for a problem where the order of the plant is greater than 3.

Speaking of the controller design, recall that $A_M = A_P - B_P F$, and since A_M, A_P, and B_P are consistent with the CCF configuration, $F = [0\ 0\ \dots\ 0\ 1](A_P - A_M) = [0.018\ -0.035\ 0.018]$.

Now it is time to "assemble" the entire model of the designed system following the previously obtained expression,

$$\begin{bmatrix} X(i+1) \\ E(i+1) \end{bmatrix} = \begin{bmatrix} A_P - B_P F & B_P F \\ \varnothing & A_P - K\hat{C} \end{bmatrix} \begin{bmatrix} X(i) \\ E(i) \end{bmatrix} + \begin{bmatrix} B_P \\ \varnothing \end{bmatrix} R(i), \ i = 1, 2, 3, \ldots$$

where $\hat{C} = [1 \ 0 \ 0 \ \ldots \ 0]$ is the matrix reflecting the fact that filter H(z) has been placed in the input of the controlled plant, also

$$A^{CL} = \begin{bmatrix} A - BF & BF \\ \varnothing & A - KC \end{bmatrix} =$$

$$\begin{bmatrix} 0 & 1 & 0 & 0 & 0 & 0 \\ 0 & 0 & 1 & 0 & 0 & 0 \\ 0.972 & -2.945 & 2.972 & 0.018 & -0.035 & 0.018 \\ 0 & 0 & 0 & -0.225 & 1 & 0 \\ 0 & 0 & 0 & -0.241 & 0 & 1 \\ 0 & 0 & 0 & 0.732 & -2.980 & 2.990 \end{bmatrix}$$

and

$$B^{CL} = \begin{bmatrix} B_P \\ \varnothing \end{bmatrix} = \begin{bmatrix} 0 \\ 0 \\ 1 \\ 0 \\ 0 \\ 0 \end{bmatrix}$$

Then the closed-loop transfer function for the reference channel of the closed-loop system, obtained using the CC software tool is

$$G^{CL}{}_R(z) = [1 \ 0 \ \ldots \ 0](zI - A^{CL})^{-1} B^{CL} = \frac{1}{z^3 - 2.972z^2 + 2.945z - .972}$$

One can realize the denominator of this expression is consistent with the denominator of the model transfer function, and with the filter

$$W(z) = 1.464e - 7(z^2 + 3.972z + .986)$$

$$= \frac{1.464e - 7(1 + 3.972z^{-1} + .986z^{-2})}{z^{-2}}$$

in the reference channel the transfer function of the resultant system is equal to the model transfer function that guarantees the full compliance with the design specifications, except the steady-state error requirement.

$$G^{CL}{}_R(z) = [1 \ 0 \ \ldots \ 0](zI - A^{CL})^{-1}B^{CL}W(z)$$
$$= \frac{1.464e - 7(z^2 + 3.972z + .986)}{z^3 - 2.972z^2 + 2.945z - .972}$$

Let us compute the steady-state error in the designed system caused by the unit step disturbance. It was shown that the error can be calculated as follows

$$M(z) = \frac{.018z^2 - .031z + .014}{z^3 - 2.765z^2 + 2.548z - .783}$$

$$N(z) = \frac{7.415e - 06(z - .993)(z - 1.00)}{z^3 - 2.765z^2 + 2.548z - .783}$$

Then the transfer function of the error channel is

$$G^D{}_{CL}(z) = \frac{G_P(z)[1 + M(z)]}{1 + M(z) + G_P(z)H(z)N(z)}$$

In our problem

$$G^D{}_{CL}(z) = \frac{0.01499(s - 0.8791)(s^2 - 1.868s + 0.8745)(s^2 - 1.992s + 0.9917)}{(s - 0.9141)(s - 0.9228)(s - 0.928)(s - 0.9802)(s^2 - 1.992s + 0.9921)}$$

and $E_{SS} = G^D{}_{CL}(1) = 2.36$ (units).

It can be seen that the error requirement has not been met. Let us investigate the opportunities for manipulating the steady state error without affecting the settling time and overshoot requirements. The results of this numerical study are summarized in the following table.

Time step	Non-dominant pole of $G_M(s)$	E_{SS}
0.005	−4	2.36
0.005	−40	1.20
0.005	−100	1.12
0.005	−200	1.10
0.002	−4	2.32
0.002	−40	1.17
0.002	−100	1.09
0.01	−4	2.41

The analysis of this table indicates that:

- an increase of the absolute value of the non-dominant pole results in the decrease of steady-state error
- a decrease of the time step results in the decrease of the steady-state error

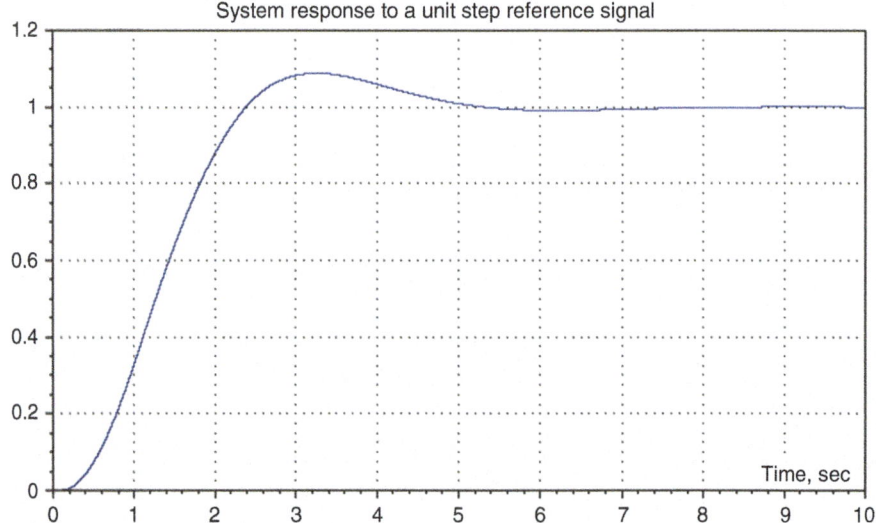

Fig. 3.16 Closed loop step response

Note that the availability of transfer functions M(z) and N(z) provides an additional opportunity to define the closed-loop transfer function $G^R_{CL}(z)$:

$$G^R_{CL}(z) = \frac{G_P(z)H(z)W(z)}{1 + M(z) + G_P(z)H(z)N(z)}$$

This definition could be preferable from the computational point of view.

The closed-loop system response is shown below in Fig. 3.16.

It should be noted that the design and especially re-design of a control system should be facilitated by a special MATLAB or in our case Program CC code:

```
tstep=.005
pole=40
format long zeros compact
gps=(3*s^2+5*s+10)/(s^3+2*s^2+3*s+5)
gp=convert(gps,8,tstep)
gms=1.33^2*pole/(s^2+2*.6*1.33*s+1.33^2)/(s+pole)
gm=convert(gms,8, tstep)
ges=1/(s+15)/(s+16)/(s+18)
ge=convert(ges,8, tstep)
geccf=ccf(ge)
(ae,be,ce,d)=unpack(geccf)
gpccf=ccf(gp)
(ap,bp,cp,d)=unpack(gpccf)
h=1/(cp(1,3)*z^2+cp(1,2)*z+cp(1,1))
```

```
gmccf=ccf(gm)
gmccf=ccf(gm);
(am,bm,cm,d)=unpack(gmccf)
w=cm(1,3)*z^2+cm(1,2)*z+cm(1,1)
f=(0,0,1)*(ap-am);
c=(1,0,0);
k(1,1)=-ae(3,3)+ap(3,3)
k(2,1)=-ae(3,2)+ap(3,2)+ap(3,3)*k(1,1)
k(3,1)=-ae(3,1)+ap(3,1)+ap(3,2)*k(1,1)+ap(3,3)*k(2,1)
m=f*(z*iden(3)-ap+k*c)^-1*bp;
n=f*(z*iden(3)-ap+k*c)^-1*k;
mm=1|m
gcl=w*(h*gp*mm)|n
gcl=near(gcl)
gerr=gp|(mm*h*n)
gerr=near(gerr)
gerr=ccf(gerr)
(ar,br,cr,d)=unpack(gerr)
err=cr*(iden(6)-ar)^-1*br
err
```

3.6   Decoupling Control of a MIMO Process

A multi-input-multi-output (MIMO) process is characterized by a transfer matrix, $G(z)$, that may comprise of several regression equations,

$$\begin{bmatrix} Y_1(z) \\ \ldots \\ Y_m(z) \end{bmatrix} = \begin{bmatrix} G_{11}(z) & \ldots & G_{1m}(z) \\ \ldots & \ldots & \ldots \\ G_{m1}(z) & \ldots & G_{mm}(z) \end{bmatrix} \begin{bmatrix} U_1(z) \\ \ldots \\ U_m(z) \end{bmatrix} \quad \text{or simply } Y(z) = G(z)U(z)$$

where notations are obvious. It is said that particular elements of the transfer matrix represent particular dynamic channels of the MIMO system, i.e.

$$G_{jk}(z) = \frac{Y_j(z)}{U_k(z)}$$

Note that the number of control efforts is expected to be equal to the number of controlled variables.

The control task of a MIMO process is complicated because of the cross-coupling effects, i.e. generally speaking any input signal affects all controlled variables and the independent access to any desired output variable can be achieved by simultaneous utilization of all input signals. The purpose of decoupling control is to enable the process operator to independently manipulate any output variable of choice without affecting other outputs. We will consider several discrete-time decoupling control techniques: steady-state decoupling, state-variable decoupling and full decoupling.

3.6.1 Steady-State Decoupling

Steady-state decoupling is an open-loop control technique that results in the independent access to the steady-state values of the output variables. While the transient process affects all output variables, the residual (i.e. steady-state) effects can be found only in the output variable(s) of choice.

Assume that transfer matrix $G(z)$ describes a combination of interrelated industrial control systems that already have been stabilized by the introduction of "local" controllers, i.e. the stability, settling times, and overshoot properties of the particular channels have been assured.

Define the input vector $U(z)$ as $W*R_0(z)$, where $R_0(z)$ is the set point vector representing the desired steady-state values of all output variables, and W is an $m \times m$ matrix gain, defined as

$$W = G(1)^{-1}$$

Indeed,

$$Y_{SS} = \lim_{k \to \infty} Y(k) = \lim_{z \to 1} \frac{z-1}{z} Y(z) = \lim_{z \to 1} \frac{z-1}{z} G(z)U(z)$$

Assume that Y^{DES} is the vector of desired steady-state values of the output variables, and define the set point $R_0(t) = Y^{DES}u(t)$ as a set of step functions of the appropriate magnitudes,

$$Y^{DES} = \begin{bmatrix} Y_1^{DES} \\ \ldots \\ Y_m^{DES} \end{bmatrix}$$

Then $R_0(z) = Y^{DES}\frac{z}{z-1}$ and consequently,

$$U(z) = WR_0(z) = \frac{z}{z-1} G^{-1}(1)Y^{DES}$$

In this case, providing that matrix $G(1)$ is not singular,

$$Y_{SS} = \lim_{k \to \infty} Y(k) = \lim_{z \to 1} \frac{z-1}{z} G(z)\frac{z}{z-1} G^{-1}(1)Y^{DES} = Y^{DES}$$

Example 3.7 Given a z-domain mathematical model of several interrelated control loops of a MIMO process relating the output variables to the input signals U_1, U_2, U_3 (time step $= 0.001$ s):

$$
\begin{bmatrix} Y_1(z) \\ Y_2(z) \\ Y_3(z) \end{bmatrix} =
\begin{bmatrix}
\dfrac{0.0495(z-0.99)}{z^2-1.97z+0.9704} & \dfrac{0.003986(z-0.995)}{z^2-1.985z+0.99} & \dfrac{0.001015(z-0.9418)}{z^2-1.97z+0.9704} \\[3mm]
\dfrac{0.00398(z-0.9802)}{(z-0.9962)(z-0.9742)} & \dfrac{0.02029(z-0.9512)}{z^2-1.979z+0.9802} & \dfrac{0.002996(z-0.9417)}{z^2-1.934z+0.9418} \\[3mm]
\dfrac{0.007884(z-0.9988)}{z^2-1.969z+0.9704} & \dfrac{0.0009998(z-0.99)}{z^2-1.989z+0.99} & \dfrac{0.05824}{z-0.9418}
\end{bmatrix}
$$
$$
\times \begin{bmatrix} U_1(z) \\ U_2(z) \\ U_3(z) \end{bmatrix}
$$

First, let us observe the cross-coupling "in action." The graph in Fig. 3.17 features the response in each of the three outputs of the system to a unit step signal applied to the input #1. It is important to observe the residual (steady-state) effects of this input signal in all outputs.

The following simulation features the step response of the same system driven through a decoupling filter assuming that R_{02} is a step with magnitude of 3 (units) and $R_{01} = R_{03} = 0$

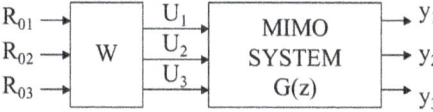

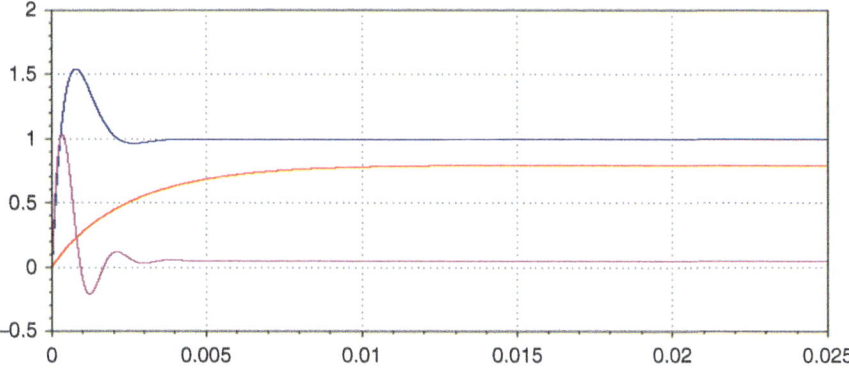

Fig. 3.17 Demonstration of cross-coupling to a step response

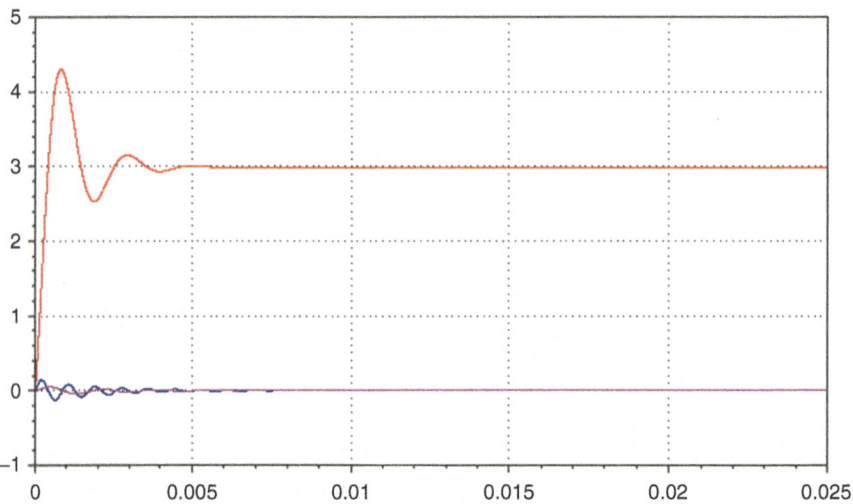

Fig. 3.18 Steady state decoupling for a step response

where
$$W = G(1)^{-1} = \begin{bmatrix} 1.0040123 & -3.113591e - 003 & -0.0752309 \\ 1.199517e - 003 & 1.0001570 & -0.0225935 \\ -0.0535559 & -6.977921e - 003 & 1.0041737 \end{bmatrix}$$

It can be seen that the goal of the steady-state decoupling control has been achieved: the intended output has been incremented by 3 units and no residual effects in other outputs are observed (Fig. 3.18).

3.6.2   Full Decoupling

The full decoupling problem does not have an assured general solution. Its solution can be worked out on a case-by-case basis as follows.

Given an $n \times n$ z-domain transfer matrix of a MIMO process, $G(z)$. Given design specifications, settling time, overshoot of the step response, and steady-state error for its direct input–output channels, i.e. $R_{0j} \rightarrow Y_j$, $j = 1,2,\ldots,n$.

The design procedure requires that a rational decoupling filter, $W(z)$, be designed that matrix $Q(z) = G(z)W(z)$ be diagonal, however, each of its diagonal elements, $Q_{jj}(z)$, should have the order of the numerator lower than the order of the denominator and does not have zeros outside the unit circle. Then diagonal elements $Q_{jj}(z)$ should be treated as transfer functions of independent single-input-single-output processes and equipped with the feedback controllers and input filters to comply with the design specifications.

Let us consider the design procedure in more detail for $n = 3$. Assume

$$G(z) = \begin{bmatrix} G_{11}(z) & G_{12}(z) & G_{13}(z) \\ G_{21}(z) & G_{22}(z) & G_{23}(z) \\ G_{31}(z) & G_{32}(z) & G_{33}(z) \end{bmatrix}$$

and

$$W(z) = \begin{bmatrix} W_{11}(z) & W_{12}(z) & W_{13}(z) \\ W_{21}(z) & W_{22}(z) & W_{23}(z) \\ W_{31}(z) & W_{32}(z) & W_{33}(z) \end{bmatrix}$$

Define elements of the matrix $W(z)$ from the following equations reflecting the requirement that matrix $Q(z) = G(z)W(z)$ must be diagonal:

$$Q_{12}(z) = G_{11}(z)W_{12}(z) + G_{12}(z)W_{22}(z) + G_{13}(z)W_{32}(z) = 0$$
$$Q_{13}(z) = G_{11}(z)W_{13}(z) + G_{12}(z)W_{23}(z) + G_{13}(z)W_{33}(z) = 0$$
$$Q_{21}(z) = G_{21}(z)W_{11}(z) + G_{22}(z)W_{21}(z) + G_{23}(z)W_{31}(z) = 0$$
$$Q_{23}(z) = G_{21}(z)W_{13}(z) + G_{22}(z)W_{23}(z) + G_{23}(z)W_{33}(z) = 0$$
$$Q_{31}(z) = G_{31}(z)W_{11}(z) + G_{32}(z)W_{21}(z) + G_{33}(z)W_{31}(z) = 0$$
$$Q_{32}(z) = G_{31}(z)W_{12}(z) + G_{32}(z)W_{22}(z) + G_{33}(z)W_{32}(z) = 0$$

Note that the above system contains six equations and nine unknowns and therefore allows for many solutions thus providing the opportunity for a rational choice of the elements of matrix $W(z)$. Then, the non-zero diagonal elements of matrix $Q(z)$ can be defined as follows

$$Q_{11}(z) = G_{11}(z)W_{11}(z) + G_{12}(z)W_{21}(z) + G_{13}(z)W_{31}(z)$$
$$Q_{22}(z) = G_{21}(z)W_{12}(z) + G_{22}(z)W_{22}(z) + G_{23}(z)W_{32}(z)$$
$$Q_{33}(z) = G_{31}(z)W_{13}(z) + G_{32}(z)W_{23}(z) + G_{33}(z)W_{33}(z)$$

In the situation when transfer functions $Q_{11}(z)$, $Q_{22}(z)$ and $Q_{33}(z)$ are acceptable, the design of three "local" control loops should follow the previously established procedures.

Note that design of a decoupling filter $W(z)$ has a more rigorous but not necessarily more simple solution: transfer matrix $W(z)$ could be defined as the adjoint matrix of $G(z)$:

$$W(z) = \text{Adj}\{G(z)\}$$

Indeed,

$$G(z)^{-1} = \frac{1}{Det[G(z)]} Adj\{G(z)\}$$

where Det[.] is the symbol of determinant. Since $G(z)G(z)^{-1} = I$, matrix multiplication

$$Q(z) = G(z) \cdot Adj\{G(z)\}$$

results in a diagonal matrix where every diagonal element is equal to the determinant Det[G(z)], i.e. $Q_{kk}(z) = Det[G(z)]$ for all $k = 1,2,\ldots$

Example 3.8 Given transfer matrix of a two-input-two-output controlled plant defined for the time step of 0.01 s:

$$G(z) = \begin{bmatrix} \dfrac{0.01995(z - 0.9753)}{z^2 - 1.97z + 0.9704} & \dfrac{0.02(z - 0.99)}{z^2 - 1.989z + 0.99} \\[4mm] \dfrac{0.01972(z - 0.998)}{(z - 0.9962)(z - 0.9742)} & \dfrac{0.0199}{z - 0.99} \end{bmatrix}$$

Define the decoupling filter

$$W(z) = \begin{bmatrix} W_{11}(z) & W_{12}(z) \\ W_{21}(z) & W_{22}(z) \end{bmatrix},$$

then elements $W_{ij}(z)$, $i,j = 1,2$, can be defined from the equations

$$\frac{0.01995(z - 0.9753)}{z^2 - 1.97z + 0.9704} W_{12}(z) + \frac{0.02(z - 0.99)}{z^2 - 1.989z + 0.99} W_{22}(z) = 0$$

$$\frac{0.01972(z - 0.998)}{(z - 0.9962)(z - 0.9742)} W_{11}(z) + \frac{0.0199}{z - 0.99} W_{21}(z) = 0$$

Assume $W_{12}(z) = G_{11}(z)^{-1}G_{12}(z)$, $W_{22}(z) = -1$, $W_{21}(z) = G_{22}(z)^{-1}G_{21}(z)$, and $W_{11}(z) = -1$, this choice results in off-diagonal elements of matrix Q(z), $Q_{12}(z) = Q_{21}(z) = 0$, and diagonal elements

$$Q_{11}(z) = \frac{-0.0001329(z^2 - 1.897z + 0.8998)(z - 0.9789)(z + 1.016)}{(z - 0.9742)(z^2 - 1.97z + 0.9704)(z^2 - 1.989z + 0.99)}$$

$$Q_{22}(z) = \frac{-0.0001326(z^2 - 1.897z + 0.8998)(z - 0.9789)(z + 1.016)}{(z - 0.9742)(z - 0.9753)(z - 0.99)(z^2 - 1.989z + 0.99)}$$

It can be seen that the choice of transfer matrix W(z) has been a success: the diagonal elements of matrix $Q(z) = G(z)W(z)$ have the required properties. Now the MIMO controlled plant with the transfer matrix G(z) driven through the decoupling

Fig. 3.19 Full decoupling
system setup

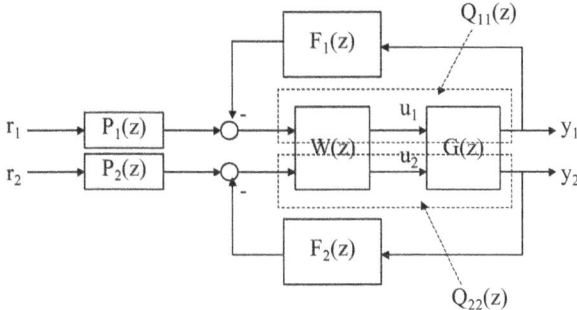

filter $W(z)$ for all practical purposes is just a combination of two completely
independent single-input-single-output systems with transfer functions $Q_{11}(z)$ and
$Q_{22}(z)$ that has to be stabilized to comply with the design specifications by introducing feedback controllers and prefilters as shown in Fig. 3.19.

Let us consider the design of the controller and prefilter for the first dynamic
channel, $Q_{11}(z)$. Assume the required settling time $T_{SET} = 5$ s, overshoot of the step
response P% <3 % and time step $=0.01$ s.

First, introduce an s-domain model transfer function, compliant with the design
specifications:

$$G_M(s) = \frac{2419}{(s+0.8)(s+3.2)(s+4.5)(s+5)(s+6)(s+7)}$$

that has the following z-domain ZOH equivalent for the time step of 0.01 s:

$$G_M(z) = \frac{1e - 10\left(0.032z^5 + 1.776z^4 + 9.060z^3 + 8.723z^2 + 1.585z + 0.027\right)}{z^6 - 5.742z^5 + 13.74z^4 - 17.52z^3 + 12.57z^2 - 4.812z + 0.767}$$

Then zeros of the transfer function $Q_{11}(z)$ will be cancelled by placing the filter

$$K(z) = -\frac{1}{0.0001329(z^2 - 1.897z + 0.8998)(z - 0.9789)(z + 1.016)}$$

in the input of the "controlled plant" and a negative feedback polynomial $H(z)$ will
be introduced such that the closed-loop system transfer function become equal to

$$\frac{Q_{11}(z)K(z)}{1 + H(z)Q_{11}(z)K(z)} = \frac{1}{H(z) + D_Q(z)}$$

where $D_Q(z)$ is the denominator of the transfer function $Q_{11}(z)$. Note that

$$\frac{1}{H(z) + D_Q(z)} = \frac{1}{H(z) + (z - 0.9742)(z^2 - 1.97z + 0.9704)(z^2 - 1.989z + 0.99)}$$
$$= \frac{1}{H(z) + z^6 - 5.929z^5 + 14.65z^4 - 19.30z^3 + 14.31z^2 - 5.659z + 0.932}$$

Recall that it is our goal to assure that the above transfer function must be modified to match the model transfer function $G_M(z)$. This modification includes,

1. Choosing polynomial $H(z)$ as

$$\begin{aligned}
H(z) &= D_M(z) - D_Q(z) \\
&= \left[z^6 - 5.742z^5 + 13.74z^4 - 17.52z^3 + 12.57z^2 - 4.812z + 0.767\right] \\
&\quad - \left[z^6 - 5.929z^5 + 14.65z^4 - 19.30z^3 + 14.31z^2 - 5.659z + 0.932\right] \\
&= 0.187z^5 - 0.912z^4 + 1.779z^3 - 1.736z^2 + 0.847z - 0.165
\end{aligned}$$

where $D_M(z)$ is the denominator of the model transfer function

2. Defining the feedback $F_1(z)$ (see the block diagram above) as

$$F_1(z) = K(z)H(z)$$
$$= \frac{0.187z^5 - 0.912z^4 + 1.779z^3 - 1.736z^2 + 0.847z - 0.165}{0.0001329(z^2 - 1.897z + 0.8998)(z - 0.9789)(z + 1.016)}$$

3. Defining the prefilter $P_1(z)$ (see the block diagram above) as

$$P_1(z) = K(z)N_M(z)$$
$$= -\frac{1e - 10(0.032z^5 + 1.776z^4 + 9.060z^3 + 8.723z^2 + 1.585z + 0.027)}{0.0001329(z^2 - 1.897z + 0.8998)(z - 0.9789)(z + 1.016)}$$

Example 3.9

Given transfer matrix of a two-input-two-output continuous-time controlled process:

$$G(s) = \begin{bmatrix} \dfrac{2}{s+1} & \dfrac{0.2}{(s+1)(s+0.1)} \\ \dfrac{0.1}{(s+1)(s+0.2)} & \dfrac{4}{s+1} \end{bmatrix}$$

Design a discrete-time decoupling control system operating with the time step of 0.02 s and meet the following design specifications: both channels must use a state-variable feedback controller, the settling time $T_{set} = 8$ s and overshoot of step response $\approx 0\%$

First, obtain a z-domain description of the controlled plant for the ZOH input:

$$G(z) = \begin{bmatrix} \dfrac{3.96e - 02}{z - .9802} & \dfrac{3.971e - 05(z + .9927)}{(z - .998)(z - .98)} \\ \dfrac{1.984e - 05(z + .992)}{(z - .996)(z - .98)} & \dfrac{.0792}{z - .9802} \end{bmatrix}$$

Using a MATLAB tool obtain $W(z) = \text{Adj}[G(z)] =$

$$\begin{bmatrix} \dfrac{.0792}{z - .98} & \dfrac{-3.9708e - 05(z + .9927)}{(z - .998)(z - .98)} \\ \dfrac{-1.984e - 05(z + .992)}{(z - .98)(z - .996)} & \dfrac{3.9603e - 02}{z - .98} \end{bmatrix}$$

Now obtain $Q(z) = G(z)W(z)$. Indeed, $Q(z)$ is a diagonal matrix and its diagonal elements, $Q_{11}(z) = Q_{22}(z) = \frac{3.137e-03(z-.9956)(z-.9984)}{(z-.998)(z-.996)(z-.98)^2}$.

The next task is to establish an s-domain transfer function representing the desired dynamics of the decoupled channel and its z-domain equivalent for ZOH input and time step of 0.02 s:

$$G_M(s) = \frac{8}{(s + .2)(s + 2)(s + 4)(s + 5)}$$

$$G_M(z) = \frac{5.1e - 08(z^3 + 10.52z^2 + 10.059z + .8742)}{(z - .996)(z - .961)(z - .923)(z - .9048)}$$

Assume that zeros of transfer function $Q_{11}(z)$ are cancelled by the introduction of a special filter

$$H(z) = \frac{1}{3.137e - 03(z - .9956)(z - .9984)} = \frac{318.776}{(z - .9956)(z - .9984)}$$

Let us represent transfer function $\overline{Q}(z) = Q_{11}(z)H(z)$ by a CCF:

$$A_Q = \begin{bmatrix} 0 & 1 & 0 & 0 \\ 0 & 0 & 1 & 0 \\ 0 & 0 & 0 & 1 \\ -.955 & 3.8645 & -5.8638 & 3.9544 \end{bmatrix}, \quad B_Q = \begin{bmatrix} 0 \\ 0 \\ 0 \\ 1 \end{bmatrix}, \text{ and}$$

$$\overline{C} = [1 \quad 0 \quad 0 \quad 0]$$

Representing transfer function $G_M(z)$ by a CCF results in:

$$A_M = \begin{bmatrix} 0 & 1 & 0 & 0 \\ 0 & 0 & 1 & 0 \\ 0 & 0 & 0 & 1 \\ -.799 & 3.384 & -5.369 & 3.785 \end{bmatrix}, \; B_M = \begin{bmatrix} 0 \\ 0 \\ 0 \\ 1 \end{bmatrix}, \; \text{and}$$

$$C_M = \begin{bmatrix} 4.459 & 51.306 & 53.657 & 5.1 \end{bmatrix} \cdot e - 08$$

Now the matrix F of the state-variable controller can be defined as:

$$F = \begin{bmatrix} 0 & 0 & 0 & 1 \end{bmatrix} \cdot (A_Q - A_M) = \begin{bmatrix} -.1557 & .4808 & -.4947 & .1696 \end{bmatrix}$$

In order to design the state observer, introduce an s-domain transfer function representing the desired dynamics of the state estimation process and its z-domain equivalent for ZOH input and time step of 0.02 s:

$$G_E(s) = \frac{1}{(s+5)(s+6)(s+7)(s+8)}$$

$$G_E(z) = \frac{6.e - 09(z^3 + 9.92z^2 + 8.94z + .732)}{(z - .904)(z - .887)(z - .869)(z - .852)}$$

The appropriate discrete-time domain fundamental matrix is:

$$A_E = \begin{bmatrix} 0 & 1 & 0 & 0 \\ 0 & 0 & 1 & 0 \\ 0 & 0 & 0 & 1 \\ -.594 & 2.709 & -4.628 & 3.513 \end{bmatrix}$$

This shall be followed by the evaluation of the element of matrix K of the state observer using the recursive formula of page 141:

$$K_1 = -A_E(4,4) + A_Q(4,4) = .4411$$
$$K_2 = -A_E(4,3) + A_Q(4,3) + A_Q(4,4)*K_1 = .5085$$
$$K_3 = -A_E(4,2) + A_Q(4,2) + A_Q(4,3)*K_1 + A_Q(4,4)*K_2 = .5794$$
$$K_4 = -A_E(4,1) + A_Q(4,1) + A_Q(4,2)*K_1 + A_Q(4,3)*K_2 + A_Q(4,4)*K_1 = .6541$$

Finally, matrix K of the state observer and its fundamental matrix M are:

$$K = \begin{bmatrix} .4411 \\ .5085 \\ .5794 \\ .6541 \end{bmatrix} \text{and } M = A_Q - K\bar{C} = \begin{bmatrix} -.4411 & 1 & 0 & 0 \\ -.5085 & 0 & 1 & 0 \\ -.5794 & 0 & 0 & 1 \\ -1.6091 & 3.8644 & -5.8638 & 3.9544 \end{bmatrix}$$

The verification of the state-variable controller could be achieved by obtaining the following closed-loop transfer function:

$$G_{CL}(z) = \overline{C} \cdot (z \cdot I - A_Q + B_Q \cdot F)^{-1} \cdot B_Q$$

$$= \frac{1}{(z - .996)(z - .961)(z - .923)(z - .9048)}$$

This indicates that the design goal is partially achieved, i.e. the denominator of the transfer function describing the dynamics of the control loop is indeed equal to the denominator of the model transfer function. The desired numerator could be obtained by placing a special filter in the reference channel of the system:

$$P(z) = \frac{5.1e - 08(z^3 + 10.52z^2 + 10.059z + .8742)}{z^4}$$

Note that having z^4 in the denominator does not affect dynamics of the control loop but may be required for the implementation of this filter.

It should be noted that since decoupled channels have identical dynamics, the obtained state-variable controller and state observer shall be utilized in both channels. The resultant designed system should be subjected to simulation analysis.

3.6.3 State-Variable Decoupling

Consider a discrete-time state-variable description of a MIMO controlled plant, controller, and a filter,

$$X(k + 1) = AX(k) + BU(k)$$
$$U(k) = WR(k) - FX(k)$$
$$Y(k) = CX(k), \quad k = 1, 2, 3, \ldots$$

Assume that X, Y, U, and R are $(n \times 1)$ vectors, A, B, W, F, and C are $(n \times n)$ matrices. This design is based on the attempt to "force" the above system to behave as a set of fully independent first-order systems with the pre-specified settling time. The problem may or may not have a solution.

First, obtain the closed-loop system description

$$X(k + 1) = AX(k) + B[WR(k) - FX(k)] \text{ or}$$
$$X(k + 1) = (A - BF)X(k) + BWR(k) \text{ or}$$
$$Y(k + 1) = CX(k + 1) = (CA - CBF)X(k) + CBWR(k) \quad k = 1, 2, 3, \ldots$$

Define the required first order dynamics of the input–output channels as

$$\frac{Y_j(s)}{R_j(s)} = \frac{\alpha_j}{s + \alpha_j}$$

where $\alpha_j > 0$ are found from the condition $T_j^{SET} = \frac{4}{\alpha_j}$ where T_j^{SET}, $j = 1,2,\ldots,n$ are the required settling times of the particular input–output channels. These first order transfer functions are to be converted into the z-domain for the particular time step of interest and using the ZOH equivalence:

$$\frac{Y_j(z)}{R_j(z)} = \frac{p_j}{z + q_j}$$

that has the following discrete-time domain representation

$$Y_j(k + 1) = -q_j Y_j(k) + p_j R_j(k), \, j = 1, 2, \ldots, n, \text{ and } k = 1, 2, 3, \ldots$$

Combining these equations results in the following matrix–vector equation

$$Y(k + 1) = QY(k) + PR(k) \text{ or } Y(k + 1) = QCX(k) + PR(k)$$

where

$$Q = \begin{bmatrix} -q_1 & 0 & \cdots & 0 \\ 0 & -q_2 & \cdots & 0 \\ \cdots & \cdots & \cdots & \cdots \\ 0 & 0 & \cdots & -q_n \end{bmatrix} \text{ and } P = \begin{bmatrix} p_1 & 0 & \cdots & 0 \\ 0 & p_2 & \cdots & 0 \\ \cdots & \cdots & \cdots & \cdots \\ 0 & 0 & \cdots & p_n \end{bmatrix}$$

Matching the above equation to the matrix–vector equation of the closed-loop system results in the following equalities

$$QC = CA - CBF \text{ and } P = CBW$$

that provide the definition for the controller matrix and the input filter

$$F = (CB)^{-1}(CA - QC)$$
$$W = (CB)^{-1}P$$

It can be seen that the existence of the inverse of matrix CB presents the only restriction to the solution of this problem.

Example 3.10 Given discrete-time domain state-variable description of a controlled plant corresponding to the time step of 0.01 s and the ZOH application:

$$
\begin{bmatrix} X_1(k+1) \\ X_2(k+1) \\ X_3(k+1) \end{bmatrix} = \begin{bmatrix} 1.0099947 & 0.0205451 & 0.0295186 \\ -0.0199078 & 1.0003840 & 0.0383450 \\ 9.414670e-03 & 0.0290774 & 0.9331087 \end{bmatrix} \begin{bmatrix} X_1(k) \\ X_2(k) \\ X_3(k) \end{bmatrix}
$$

$$
+ \begin{bmatrix} 0.0209903 & 3.520221e-04 & 0.0195443 \\ 9.673052e-04 & 0.0201970 & -0.0403990 \\ 0.0580587 & 9.954242e-03 & -0.0101520 \end{bmatrix} \begin{bmatrix} U_1(k) \\ U_2(k) \\ U_3(k) \end{bmatrix}
$$

and

$$
\begin{bmatrix} Y_1(k) \\ Y_2(k) \\ Y_3(k) \end{bmatrix} = \begin{bmatrix} 1 & 1 & 2 \\ 2 & 1 & 5 \\ 5 & -2 & 0 \end{bmatrix} \begin{bmatrix} X_1(k) \\ X_2(k) \\ X_3(k) \end{bmatrix}
$$

The design specifications call for the overshoot under 3 % and settling times of 2 s, 5 s and 1 s for the respective decoupled channels of the closed-loop system.

First, express these requirements in terms of three s-domain first order transfer functions and their appropriate z-domain equivalents:

$$
\frac{Y_1(s)}{R_1(s)} = \frac{2}{s+2} \quad \text{and} \quad \frac{Y_1(z)}{R_1(z)} = \frac{1.98e-02}{z-.980}
$$

$$
\frac{Y_2(s)}{R_2(s)} = \frac{0.8}{s+0.8} \quad \text{and} \quad \frac{Y_2(z)}{R_2(z)} = \frac{7.968e-03}{z-.992}
$$

$$
\frac{Y_3(s)}{R_3(s)} = \frac{4}{s+4} \quad \text{and} \quad \frac{Y_3(z)}{R_3(z)} = \frac{3.921e-02}{z-.961}
$$

Consequently, matrices Q and P are as follows:

$$
Q = \begin{bmatrix} 0.980 & 0 & 0 \\ 0 & 0.992 & 0 \\ 0 & 0 & 0.961 \end{bmatrix} \quad \text{and} \quad P = \begin{bmatrix} 0.0198 & 0 & 0 \\ 0 & 7.968e-03 & 0 \\ 0 & 0 & 0.03921 \end{bmatrix}
$$

and finally,

$$
W = (CB)^{-1}P = \begin{bmatrix} -6.172941e-03 & 1.312376e-03 & -9.225882e-04 \\ 0.0385518 & -6.843106e-03 & 7.611353e-03 \\ 0.0119965 & -2.249788e-03 & 4.382294e-03 \end{bmatrix}
$$

and

$$
F = (CB)^{-1}(CA - QC) = \begin{bmatrix} 0.3918118 & 0.4969176 & -2.3058824 \\ -1.7369471 & 0.5169294 & 9.6435294 \\ -0.3706059 & 0.5011412 & 3.8129412 \end{bmatrix}
$$

The analysis of the closed-loop transfer matrix

$$G_{CL}(z) = C(zI - A + BF)^{-1}BW$$

indicates that $G_{CL}(z)$ is indeed a diagonal matrix and its diagonal elements are

$$\frac{1.98e - 02}{z - .980}, \quad \frac{7.968e - 03}{z - .992}, \quad \frac{3.921e - 02}{z - .961}$$

Exercise 3.2

Problem 1 The controlled plant is defined by its transfer function

$$G_P(s) = \frac{5s^2 + 4s + 1}{s^3 + 2s^2 + 3s + 10}$$

Design a discrete-time state-variable controller with a discrete-time state observer operating with the time step of 0.005 s. The design specifications call for the settling time of 5 s, overshoot of step response under 3 %, and the steady-state error under 0.2 (units) for a unit step disturbance signal.

Problem 2 Given the transfer matrix of a controlled plant:

$$G_P(s) = \begin{bmatrix} \dfrac{s+2}{s^2+s+6} & \dfrac{.5}{s+8} \\[3mm] \dfrac{.3}{s^2+s+25} & \dfrac{s+10}{s^2+8s+12} \end{bmatrix}$$

Design a discrete-time steady-state decoupling system operating with the time step of 0.01 s. Verify your design by computer simulation. It is required that each dynamic channel has settling time of 2 s and the overshoot of step response under 3 %

Problem 3 Given continuous-time domain state equations of a controlled plant

$$\begin{aligned}
\dot{x}_1 &= 4x_1 + 3x_2 + 8x_3 - 3u_1 + u_2 - u_3 \\
\dot{x}_2 &= 6x_1 - 5x_2 - 2x_3 + u_1 + 2u_2 + u_3 \\
\dot{x}_3 &= -x_1 + 7x_2 + 9x_3 - 4u_1 + u_2 - 5u_3 \\
y_1 &= -3x_1 + 7x_2 + 4x_3 \\
y_2 &= -x_1 + x_2 - 2x_3 \\
y_3 &= x_1 + 3x_2 - x_3
\end{aligned}$$

Design a discrete-time state-variable decoupling system operating with the time step of 0.01 s. The design specifications call for the settling times of 3 s, 6 s and 8 s and the overshoot under 3 % for the respective input/output channels. Show all stages of your design. Verify your results by computer simulation.

3.7 Direct Self-Tuning Control

Consider a controlled plant with the transfer function defined in the Z-domain presumably for a ZOH input,

$$G(z) = \frac{b_{n-1}z^{n-1} + b_{n-2}z^{n-2} + \ldots + b_1 z + b_0}{z^n + a_{n-1}z^{n-1} + a_{n-2}z^{n-2} + \ldots + a_1 z + b_0} = \frac{N_P(z)}{D_P(z)}$$

Assume that the desired closed-loop operation of the control system is specified by the Z-domain model transfer function that should be of the same order as the plant,

$$G_M(z) = \frac{c_L z^L + c_{L-1}z^{L-1} + \ldots + c_1 z + c_0}{z^n + d_{n-1}z^{n-1} + d_{n-2}z^{n-2} + \ldots + d_1 z + d_0} = \frac{N_M(z)}{D_M(z)}$$

where notations are self-explanatory.

For simplicity, it is also assumed that the plant transfer function $G_P(z)$ does not have zeros outside the unit circle.

Assume the following configuration of the resultant discrete-time domain control system in Fig. 3.20.

Where $\frac{N_F(z)}{D_F(z)}$ and $\frac{N_C(z)}{D_C(z)}$ represent a filter in the reference channel and the feedback controller, $R(z)$ and $U(z)$ are the reference signal and the control effort. It is expected that the order of polynomial $N_F(z)$ is equal to the "order of the controlled plant minus 1".

Recall from your feedback controller design experience that since $G_P(z)$ does not have "bad" zeros, $D_C(z) = D_F(z) = N_P(z)$ and consequently the overall transfer function of the control system is

$$G_{CL}(z) = \frac{N_F(z)}{D_F(z)} \frac{\dfrac{N_P(z)}{D_P(z)}}{1 + \dfrac{N_P(z)N_C(z)}{D_P(z)D_C(z)}} = \frac{N_F(z)}{N_P(z)} \frac{N_P(z)D_C(z)}{D_P(z)D_C(z) + N_P(z)N_C(z)}$$

$$= \frac{N_F(z)D_C(z)}{D_P(z)D_C(z) + D_C(z)N_C(z)}$$

$$= \frac{N_F(z)}{D_P(z) + N_C(z)}$$

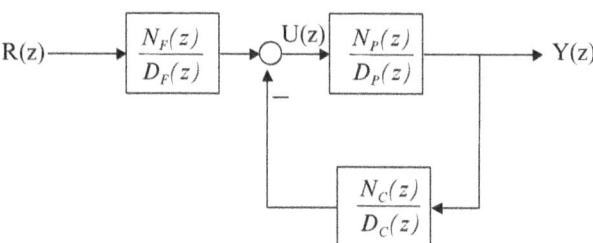

Fig. 3.20 Discrete-time domain control system

It is clear that the system would comply with the design specifications if

$$G_{CL}(z) = \frac{N_F(z)}{D_P(z) + N_C(z)} = G_M(z) = \frac{N_M(z)}{D_M(z)}$$

Then,

$$N_F(z) = N_M(z)$$

and

$$D_P(z) + N_C(z) = D_M(z)$$

Multiplying the above expression by Y(z) yields

$$Y(z)D_P(z) + Y(z)N_C(z) = Y(z)D_M(z)$$

Recall the according to the transfer function of the controlled plant, $G(z) = \frac{N_P(z)}{D_P(z)} = \frac{Y(z)}{U(z)}$, and

$$Y(z)D_P(z) = U(z)N_P(z)$$

therefore

$$U(z)N_P(z) + Y(z)N_C(z) = Y(z)D_M(z)$$

or

$$U(z)D_C(z) + Y(z)N_C(z) = Y(z)D_M(z)$$

Finally,

$$U(z)\frac{1}{D_M(z)}D_C(z) + Y(z)\frac{1}{D_M(z)}N_C(z) = Y(z)$$

Note that in the above expression $D_M(z)$ is a known polynomial and $\frac{1}{D_M(z)}$ can be viewed as a transfer function of a digital filter converting signal U(z) into signal

$$U^{NEW}(z) = U(z)\frac{1}{D_M(z)},$$

Similarly, signal

$$Y^{NEW}(z) = Y(z)\frac{1}{D_M(z)}$$

can be introduced. Then the above equation can be rewritten as

$$U^{NEW}(z)D_C(z) + Y^{NEW}(z)N_C(z) = Y(z) \tag{v}$$

This equation, converted into the discrete-time domain facilitates the direct estimation of the parameters of the controller. Note that the plant transfer function is not used in this equation. Indeed, the controller design does not require the knowledge of the mathematical description of the controlled plant! The following example provides an illustration of how this is done (for simplicity the order of the plant $n = 4$).
    Assume

$$G(z) = \frac{b_3 z^3 + b_2 z^2 + b_1 z + b_0}{z^4 + a_3 z^3 + a_2 z^2 + a_1 z + b_0} = \frac{N_P(z)}{D_P(z)}$$

and

$$G_M(z) = \frac{c_3 z^3 + c_2 z^2 + c_1 z + c_0}{z^4 + d_3 z^3 + d_2 z^2 + d_1 z + d_0} = \frac{N_M(z)}{D_M(z)}$$

Introduce

$$N_F(z) = c_3 z^3 + c_2 z^2 + c_1 z + c_0$$

$$U^{NEW}(z) = U(z) \frac{1}{z^4 + d_3 z^3 + d_2 z^2 + d_1 z + d_0}$$

$$= U(z) \frac{z^{-4}}{1 + d_3 z^{-1} + d_2 z^{-2} + d_1 z^{-3} + d_0 z^{-4}}$$

or

$$U^{NEW}(z)\left(1 + d_3 z^{-1} + d_2 z^{-2} + d_1 z^{-3} + d_0 z^{-4}\right) = U(z) z^{-4}$$

or

$$U^{NEW}(z) = U(z) z^{-4} - U^{NEW}(z)\left(d_3 z^{-1} + d_2 z^{-2} + d_1 z^{-3} + d_0 z^{-4}\right)$$

or in the discrete-time domain

$$U^{NEW}(i) = U(i-4) - d_3 U^{NEW}(i-1) - d_2 U^{NEW}(i-2) - d_1 U^{NEW}(i-3)$$
$$- d_0 U^{NEW}(i-4)$$

Similarly,

$$Y^{NEW}(i) = Y(i-4) - d_3 Y^{NEW}(i-1) - d_2 Y^{NEW}(i-2) - d_1 Y^{NEW}(i-3)$$
$$- d_0 Y^{NEW}(i-4)$$

where $i = 1,2,\ldots$ is discrete-time index.

These recursive formulae utilize known parameters of the model transfer function, measurements of the control effort and the plant output, previously calculated filtered values of the control effort and the plant output, and are easy to implement in a computer code.

Let us now return to equation ($\vee$) assuming that

$$N_C(z) = h_3 z^3 + h_2 z^2 + h_1 z + h_0$$
$$D_C(z) = N_P(z) = b_3 z^3 + b_2 z^2 + b_1 z + b_0$$

where h_3, h_2, h_1, h_0 and b_3, b_2, b_1, b_0 are unknown coefficients,

$$U^{NEW}(z)(b_3 z^3 + b_2 z^2 + b_1 z + b_0) + Y^{NEW}(z)(h_3 z^3 + h_2 z^2 + h_1 z + h_0) = Y(z)$$

Multiplying these equation by z^{-3} results in

$$U^{NEW}(z)(b_3 + b_2 z^{-1} + b_1 z^{-2} + b_0 z^{-3}) + Y^{NEW}(z)(h_3 + h_2 z^{-1} + h_1 z^{-2} + h_0 z^{-3})$$
$$= z^{-3} Y(z)$$

that can be interpreted in the discrete-time domain as

$$b_3 U^{NEW}(i) + b_2 U^{NEW}(i-1) + b_1 U^{NEW}(i-2) + b_0 U^{NEW}(i-3) + h_3 Y^{NEW}(i)$$
$$+ h_2 Y^{NEW}(i-1) + h_1 Y^{NEW}(i-2) + h_0 Y^{NEW}(i-3) = Y(i-3)$$

Introduce auxiliary variables,

$$
\begin{array}{ll}
x_1(i) = U^{NEW}(i) & x_5(i) = Y^{NEW}(i) \\
x_2(i) = U^{NEW}(i-1) & x_6(i) = Y^{NEW}(i-1) \\
x_3(i) = U^{NEW}(i-2) & x_7(i) = Y^{NEW}(i-2) \\
x_4(i) = U^{NEW}(i-3) & x_8(i) = Y^{NEW}(i-3) \\
w(i) = Y(i-3) &
\end{array}
$$

and a vector of unknown coefficients

$$\begin{bmatrix} \alpha_1 & \alpha_2 & \alpha_3 & \alpha_4 & \alpha_5 & \alpha_6 & \alpha_7 & \alpha_8 \end{bmatrix}^T = \begin{bmatrix} b_3 & b_2 & b_1 & b_0 & h_3 & h_2 & h_1 & h_0 \end{bmatrix}^T$$

Then the controller design problem can be interpreted as the estimation of the unknown vector of coefficients,

$$A = \begin{bmatrix} \alpha_1 & \alpha_2 & \alpha_3 & \alpha_4 & \alpha_5 & \alpha_6 & \alpha_7 & \alpha_8 \end{bmatrix}^T$$

using the available measurements of the "vector of input variables",

$$X(i) = \begin{bmatrix} x_1 & x_2 & x_3 & x_4 & x_5 & x_6 & x_7 & x_8 \end{bmatrix}^T$$

and the "output variable, w(i) (^{T} is the transpose symbol). One can realize that this is a typical problem that could be solved by the application of the LSM or RLSM approach.

It is good to realize that since the initial numerical values of the parameters A_0, may not be consistent with the properties of the controlled plant, application of the RLSM estimation of these parameters results in the gradual improvement of the operation of the closed-loop system thus its output variable converges to the output variable of the model transfer function $G_M(z)$.

Application of the RLSM with exponential forgetting results in a more realistic situation: parameters of the control law are being continuously adjusted in order to track time-varying properties of the controlled plant.

The block diagram in Fig. 3.21 illustrates the principle of operation of the resultant adaptive control system.

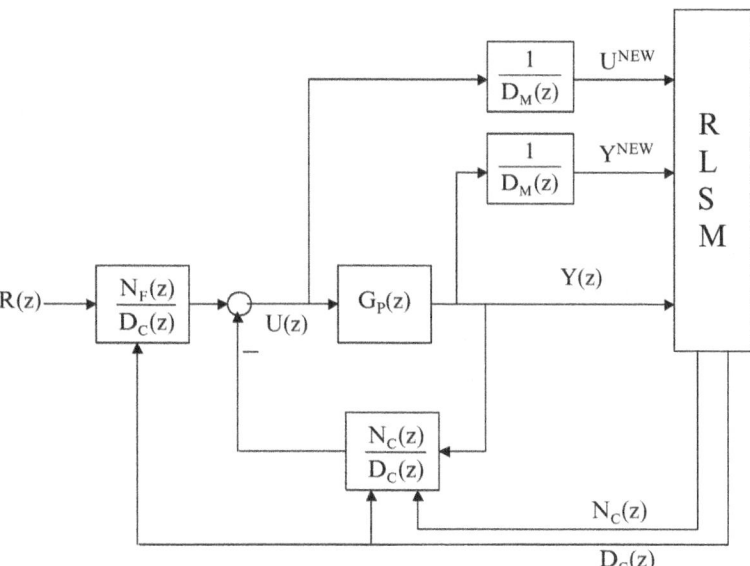

Fig. 3.21 RLSM adaptive control system

Example 3.11 The given plant is

$$G_p(s) = \frac{b_2 s^2 + b_1 s + b_0}{s^3 + a_2 s^2 + a_1 s + a_0}$$

Where as the "true" plant (assumed to be unknown to the designer) is

$$G_p(s) = \frac{s^2 + 2s + 10}{s^3 + 5s^2 + 10s + 20}$$

The desired model specifications are $T_{set} \leq 4$ s and overshoot P% ≤ 10 %. Hence we can define the dominant poles as

$$p_{1,2} = -1 \pm 1.0202j$$

Let the non-dominant pole be

$$p_3 = -10$$

Consequently, the s-domain model transfer function in continuous time domain will be

$$G_m(s) = \frac{20.41}{(s^3 + 12s^2 + 22.04s + 20.41)}$$

And the discrete time model transfer function will be

$$G_m(z) = \frac{2.564 \cdot 10^{-5} z^2 + 9.666 \cdot 10^{-5} z + 2.274 \cdot 10^{-5}}{z^3 - 2.779 z^2 + 2.565z - 0.7866}$$

The closed-loop transfer function could be found as

$$G_{CL}(z) = \frac{N_f(z)}{D_p(z) + N_c(z)} = G_m(z) = \frac{N_m(z)}{D_m(z)}$$

Then,

$$N_f(z) = N_m(z)$$

and

$$D_p(z) + N_c(z) = D_m(z)$$

Multiplying the above equation by $Y(z)$ yields

$$Y(z)D_p(z) + Y(z)N_c(z) = Y(z)D_m(z)$$

But from the system transfer function,

$$\frac{N_p(z)}{D_p(z)} = \frac{Y(z)}{U(z)}$$
$$U(z)N_p(z) = Y(z)D_p(z)$$

Then,

$$U(z)N_p(z) + Y(z)N_c(z) = Y(z)D_m(z)$$

or,

$$U(z)D_c(z) + Y(z)N_c(z) = Y(z)D_m(z)$$

Let,

$$N_c(z) = \alpha_2 z^2 + \alpha_1 z + \alpha_0$$
$$D_c(z) = b_2 z^2 + b_1 z + b_0$$

For the given problem, the above equation will be

$$U(z)(b_2 z^2 + b_1 z + b_0) + Y(z)(\alpha_2 z^2 + \alpha_1 z + \alpha_0)$$
$$= Y(z)(z^3 - 2.779 \, z^2 + 2.565 \, z - 0.7866)$$

Rearranging a few terms,

$$\left(\frac{U(z)}{z^3 - 2.779 \cdot z^2 + 2.565 \, z - 0.7866}\right)(b_2 z^2 + b_1 z + b_0)$$
$$+ \left(\frac{Y(z)}{z^3 - 2.779 \, z^2 + 2.565 \, z - 0.7866}\right)(\alpha_2 z^2 + \alpha_1 z + \alpha_0)$$
$$= Y(z)$$

Where,

$$\left(\frac{U(z)}{z^3 - 2.779 \, z^2 + 2.565 \, z - 0.7866}\right) = U^{new}(z)$$
$$\left(\frac{Y(z)}{z^3 - 2.779 \, z^2 + 2.565 \, z - 0.7866}\right) = Y^{new}(z)$$

(3.2)

are the filtered responses.

Let's call $(\alpha_2 z^2 + \alpha_1 z + \alpha_0) = (h_2 z^2 + h_1 z + h_0)$ for simplicity in utilizing same coefficients in the simulation.

Now the final expression for direct self tuning controller is

$$U^{new}(z)D_c(z) + Y^{new}(z)N_c(z) = Y(z)$$

i.e.

$$U^{new}(z)\big(b_2z^2 + b_1z + b_0\big) + Y^{new}(z)\big(h_2z^2 + h_1z + h_0\big) = Y(z) \qquad (3.3)$$

To estimate these parameters we use recursive Least Squares method (RLSM), see Sect. 2.3. It is known that as the number of RLSM iterations increases, parameter estimates converge to the solution of the LSM problem regardless of their initial values A_0 and the choice of initial matrix $P_0 = \alpha I$, where arbitrary $\alpha > 0$ determines the rate of convergence.

In our example the RLSM block, readily available in Simulink, was utilized. The forgetting factor was set as $\beta = 0.5$, parameter α of the initial matrix $P_0 = \alpha I$ was selected as $\alpha = 0.01$, the total number of iterations was limited to 1000, and initial values of estimated parameters A_0 were set to zero. The simulation setup is shown below.

As could be seen at the simulation setup above in Fig. 3.22, the following variables of the system are calculated:

output of the controlled plant

$$Y(z) = G_P(z) \cdot U(z) = \frac{0.01942z^2 - 0.038z + 0.01866}{z^3 - 2.901z^2 + 2.806z - 0.9048} \cdot U(z)$$

output of the model

$$Y_M(z) = G_M(z) \cdot R(z) = \frac{2.564 \cdot 10^{-5}z^2 + 9.666 \cdot 10^{-5}z + 2.274 \cdot 10^{-5}}{z^3 - 2.779z^2 + 2.565z - 0.7866} \cdot R(z)$$

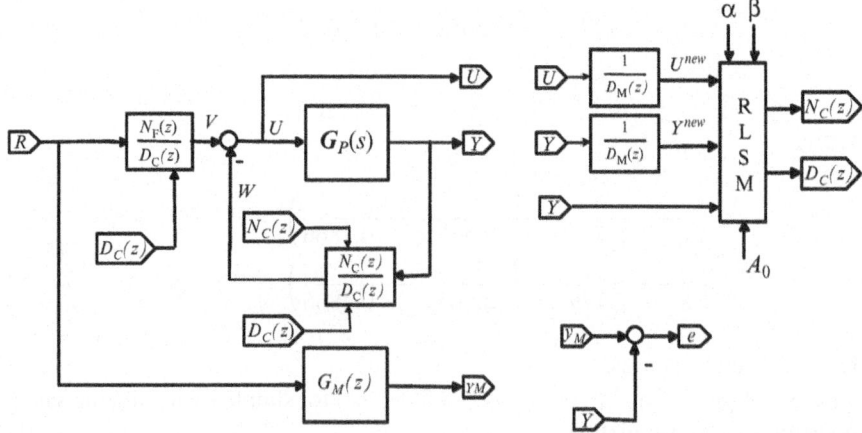

Fig. 3.22 RLSM simulation setup

output of the prefilter

$$V(z) = \frac{N_M(z)}{D_C(z)} \cdot R(z) = \frac{2.564 \cdot 10^{-5} z^2 + 9.666 \cdot 10^{-5} z + 2.274 \cdot 10^{-5}}{b_2 z^2 + b_1 z + b_0} \cdot R(z)$$

output of the controller

$$W(z) = \frac{N_c(z)}{D_c(z)} \cdot Y(z) = \frac{h_2 z^2 + h_1 z + h_0}{b_2 z^2 + b_1 z + b_0} \cdot Y(z)$$

control effort:

$$U(z) = V(z) - W(z)$$

variable $U^{new}(z)$

$$U^{new}(z) = \frac{1}{z^3 - 2.779 z^2 + 2.565 z - 0.7866} \cdot U(z)$$

variable $Y^{new}(z)$

$$Y^{new}(z) = \frac{1}{z^3 - 2.779 z^2 + 2.565 z - 0.7866} \cdot Y(z)$$

error representing the discrepancy between the self-tuning system and the model

$$e(z) = Y_M(z) - Y(z)$$

The $R(t)$ was a unit step signal

The RLSM outputs were interpreted as time-dependent parameters of the controller and prefilter, i.e.

$$A(k) = \begin{bmatrix} a_1(k) \\ a_2(k) \\ a_3(k) \\ a_4(k) \\ a_5(k) \\ a_6(k) \end{bmatrix} = \begin{bmatrix} b_2(k) \\ b_1(k) \\ b_0(k) \\ h_2(k) \\ h_1(k) \\ h_0(k) \end{bmatrix}, k = 1, 2, \ldots, 1000$$

The "final values" of the obtained parameters are

Parameters	Values
b_2	0.0194
b_1	−0.0380
b_0	0.0187
h_2	0.1222
h_1	−0.2405
h_0	0.1182

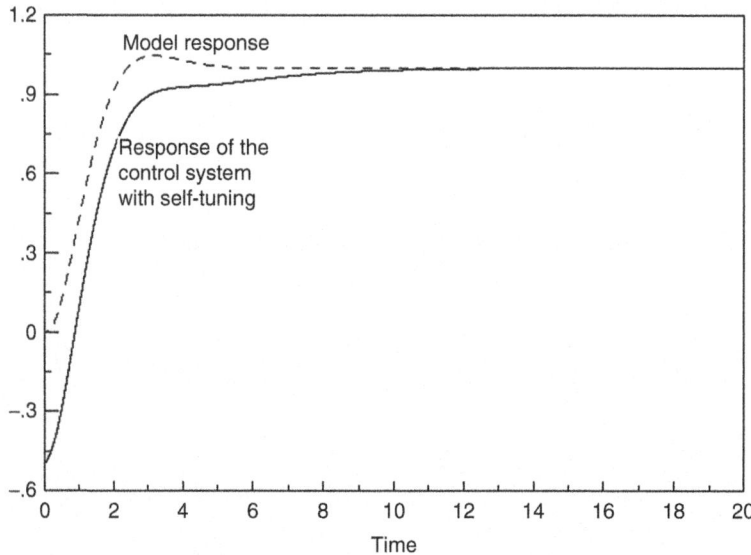

Fig. 3.23 Simulation result to step response of model and self-tuning defined system

As can be seen from Fig. 3.23, the output of the plant, driven by self-tuning controller, converges to the output of the model over 1000 steps of the RLSM procedure.

Exercise 3.3

Develop and implement in the simulation environment a discrete-time self-tuning control system under the following assumptions:

The controlled plant is a first order system that does not have right-hand-side zeros.

The "true" transfer function of the controlled plant, "unknown to the system designer", is

$$G_{TRUE}(s) = \frac{9}{s^2 + 4s + 13}$$

The design specifications are: system settling time $T_{SET} = 6$ s, overshoot of the step response $P\% \leq 2$ %, discrete time step is 0.05 s

Solutions

Exercise 3.1: Problem 1

First, a simulation setup was used to generate the input signal, $u(t)$, and simulate the output, $y(t)$, of the continuous-time dynamic system with transfer function

$$G_P(s) = \frac{s^2 + 6s + 10}{s^3 + 2s^2 + 9s + 8}$$

Over 500 samples of input–output data were taken with a time step of 0.05 s and recorded in the data array X_{500} and data array Y_{500} defined as follows:

$$X_{500} = \begin{bmatrix} -y(3) & -y(2) & -y(1) & u(3) & u(2) & u(1) \\ -y(4) & -y(3) & -y(2) & u(4) & u(3) & u(2) \\ \cdots & \cdots & \cdots & \cdots & \cdots & \cdots \\ -y(501) & -y(500) & -y(499) & u(501) & u(500) & u(499) \\ -y(502) & -y(501) & -y(500) & u(502) & u(501) & u(500) \end{bmatrix} \text{ and }$$

$$Y_{500} = \begin{bmatrix} y(4) \\ y(5) \\ \cdots \\ y(502) \\ y(503) \end{bmatrix}$$

The estimated coefficients of the z-domain transfer function

$$G_P(z) = \frac{A(4)z^2 + A(5)z + A(6)}{z^3 + A(1)z^2 + A(2)z + A(3)}$$

were obtained as

$$\overline{A} = \left(X_{500}{}^T \cdot X_{500}\right)^{-1} \left(X_{500}{}^T \cdot Y_{500}\right) = \begin{bmatrix} A(1) \\ A(2) \\ A(3) \\ A(4) \\ A(5) \\ A(6) \end{bmatrix} = \begin{bmatrix} -2.883 \\ 2.798 \\ -.905 \\ .0564 \\ -.0942 \\ .905 \end{bmatrix}$$

The following are simulated step responses of the original continuous-time system defined by transfer function $G_P(s)$ and its discrete-time model obtained from the recorded data:

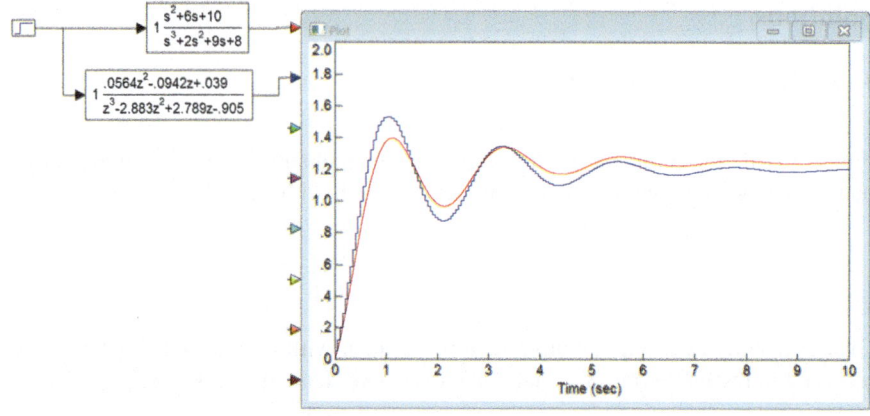

The estimated coefficients of the transfer function $G_P(z)$ were verified by obtaining this transfer function directly by conversion of the original $G_P(s)$ into the z-domain using the ZOH option and the time step of 0.05 s:

$$G_P^{DIRECT}(z) = \frac{.054847z^2 - .094252z + .040593}{z^3 - 2.88298z^2 + 2.78877z - .904837}$$

It could be seen below that step responses of the continuous-time system defined by $G_P(s)$ and its discrete-time equivalent $G_P^{DIRECT}(z)$ perfectly match:

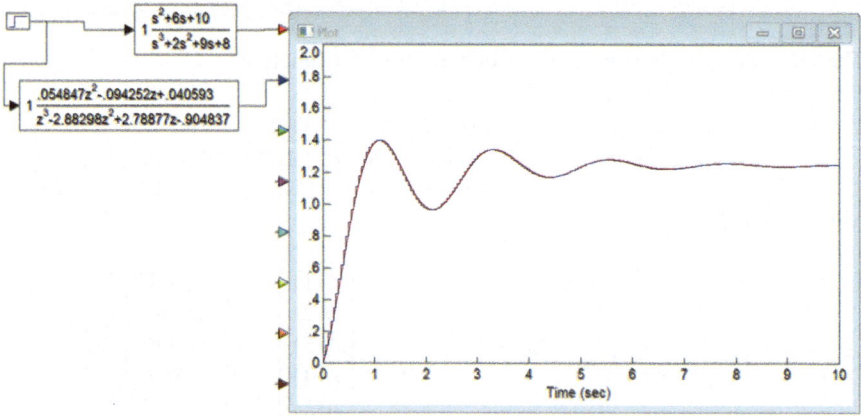

Exercise 3.1: Problem 2

Assume that the controlled plant is represented by its estimated discrete-time transfer function:

$$G_P(z) = \frac{.0564z^2 - .0942z + .039}{z^3 - 2.883z^2 + 2.789z - .905}$$

that is reasonably close to its "true" transfer function $G_P^{DIRECT}(z)$. To achieve the design specs, the desired s-domain closed-loop system transfer function has a dominant pole of $-2 \pm 2.16j$, and the non-dominant pole is chosen as 20 times the size of the dominant pole, such that the steady state errors should be sufficiently small:

$$G_M(s) = \frac{173.4}{(s + 20)(s^2 + 4s + 8.67)}$$

Conversion of this transfer function into the z-domain with the time step of 0.05 s and ZOH option results in:

$$G_M(z) = \frac{2.7168e - 03\ z^2 + 8.1852e - 3\ z + 1.4936e - 3}{z^3 - 2.167\ z^2 + 1.4806\ z - 0.3012}$$

The step response of this model, consistent with the design specifications, is shown below:

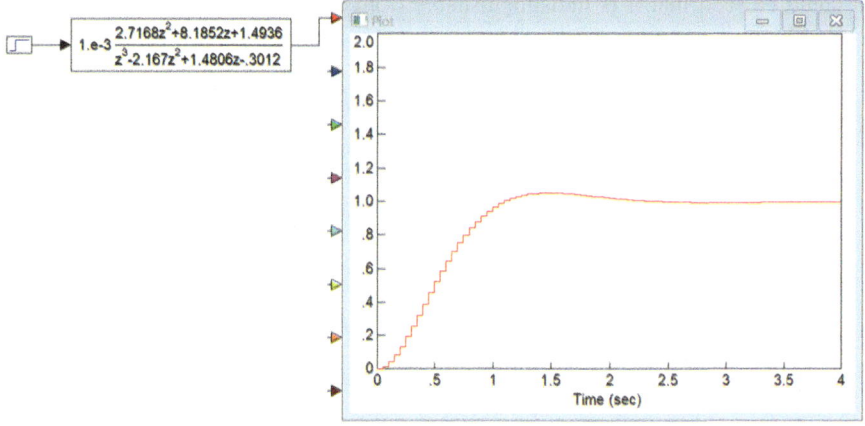

Now, the numerator of the feedback controller will be defined as the "desired closed-loop characteristic polynomial minus the characteristic polynomial of the controlled plant", i.e.

$$\left[z^3 - 2.167\ z^2 + 1.4806\ z - 0.3012\right] - \left[z^3 - 2.883z^2 + 2.789z - .905\right]$$

$$= .716z^2 - 1.3084z + .6038$$

Then the denominator of the feedback controller is defined as the numerator of the transfer function of the controlled plant, and the entire transfer function of the controller is:

$$H(z) = \frac{.716z^2 - 1.3084z + .6038}{.0564z^2 - .0942z + .039}$$

The transfer function of the filter in the reference channel, $W(z)$, is defined as "numerator of the desired closed-loop transfer function $G_M(z)$ over the numerator of the transfer function of the controlled plant", and the entire transfer function of the controller is:

$$W(z) = \frac{2.7168e - 03\ z^2 + 8.1852e - 3\ z + 1.4936e - 3}{.0564z^2 - .0942z + .039}$$

The following simulation setup demonstrates the step response of the designed system:

$$G_{CL}(z) = W(z)\frac{G_P(z)}{1 + G_P(z) \cdot H(z)}$$

It could be seen that it is indistinguishable from the step response of the desired closed-loop system $G_M(z)$

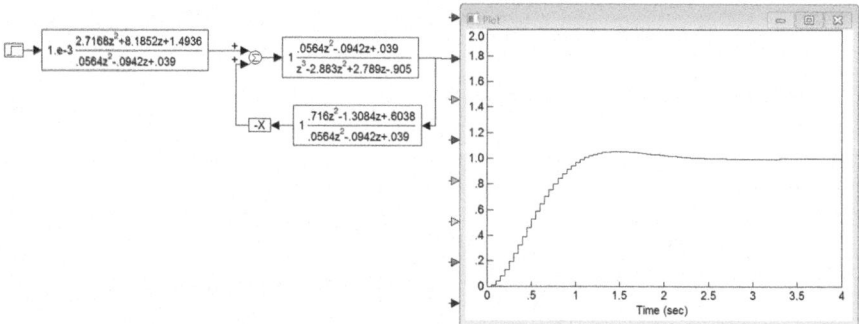

Recall that transfer function $G_P(z)$ is an imperfect discrete-time approximation of the properties of the continuous-time "true" controlled plant. Let us investigate the performance of the designed system with respect to the "true" plant:

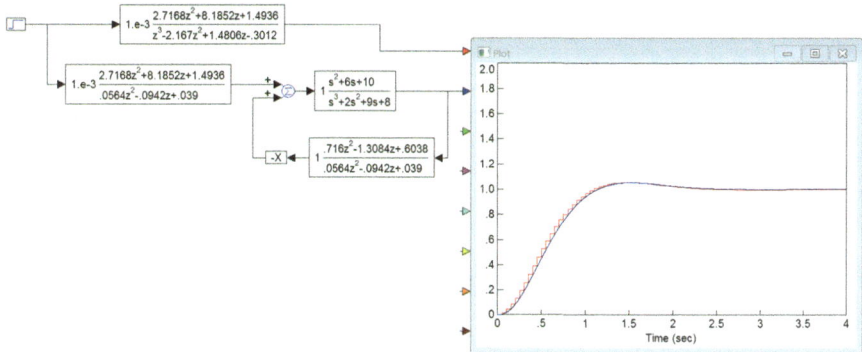

It could be seen that the designed system performance is fairly close to the desired system performance represented by transfer function $G_M(z)$.

Finally, to investigate the compliance with steady-state error requirement, a unit step disturbance signal is applied directly to the input of the controlled plant, as shown below. It could be seen that the steady-state response of the system is approximately equal to 0.1 (units) that indicates that the requirement is met. The effectiveness of the designed control circuitry is demonstrated by the next simulation: disconnection of the feedback results in fully unacceptable system operation in terms of dynamics and in terms of the sensitivity to disturbance signals.

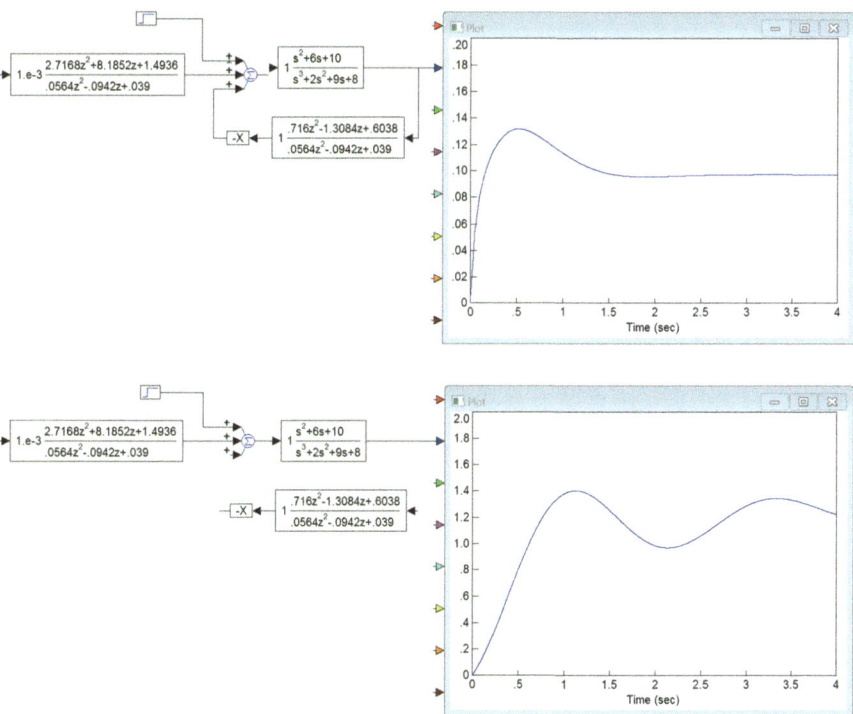

Exercise 3.1: Problem 3

Represent the given equations in a matrix–vector form

$$\dot{X} = AX + Bu, \quad y = CX$$

where

$$A = \begin{bmatrix} -10.4 & 10.3 & 8.8 \\ .6 & -.65 & -.2 \\ -11.9 & 11.7 & 9.6 \end{bmatrix}, \quad B = \begin{bmatrix} -3 \\ 1 \\ -4 \end{bmatrix}, \quad C = [-7.3 \quad 7.6 \quad 6.8]$$

Converting this state-variable form into the discrete-time domain, time step $\Delta t = 0.05$ s, using six terms of the series (note that the result is consistent with ZOH application):

$$X[k+1] = \bar{A}X[k] + \bar{B}u[k], \quad y[k] = \bar{C}X[k] = CX[k]$$

where

$$\bar{A} = \begin{bmatrix} .49205 & .5014 & .4206 \\ .0248 & .9727 & -.00574 \\ -.5744 & .5628 & 1.4615 \end{bmatrix}, \quad \bar{B} = \begin{bmatrix} -.1422 \\ .048 \\ -.1889 \end{bmatrix}, \quad C = [-7.3 \quad 7.6 \quad 6.8]$$

Exercise 3.1: Problem 4

Converting the state-variable description obtained in the above problem into canonical controllable form (CCF):

$$\bar{A}^{CCF} = P\bar{A}P^{-1}, \quad \bar{B}^{CCF} = P\bar{B}, \quad \bar{C}^{CCF} = \bar{C}P^{-1}$$

where conversion filter

$$P = \begin{bmatrix} 2.8304e3 & 1.4367e3 & -1.7654e3 \\ 2.4425e3 & 1.82306e3 & -1.3752e3 \\ 2.0370e3 & 2.224e3 & -973.46 \end{bmatrix}$$

$$\bar{A}^{CCF} = \begin{bmatrix} 0 & 1 & 0 \\ 0 & 0 & 1 \\ .93 & -2.8564 & 2.92627 \end{bmatrix}, \quad \bar{B}^{CCF} = \begin{bmatrix} 0 \\ 0 \\ 1 \end{bmatrix},$$

$$\bar{C}^{CCF} = [.10373 \quad -.22179 \quad .11821]$$

Note that the state vector of the CCF, $X^{CCF}[k] = PX[k]$, where $X[k]$ is the state vector of "real" variables.

Since design specifications in this problem are consistent with those of Problem 1, let us utilize the same model describing the required closed-loop dynamics:

$$G_M(z) = \frac{2.7168e - 03\, z^2 + 8.1852e - 3\, z + 1.4936e - 3}{z^3 - 2.167\, z^2 + 1.4806\, z - 0.3012}$$

The CCF equivalent of this model has

$$A_M = \begin{bmatrix} 0 & 1 & 0 \\ 0 & 0 & 1 \\ .3012 & -1.4806 & 2.167 \end{bmatrix}, \quad B_M = \begin{bmatrix} 0 \\ 0 \\ 1 \end{bmatrix},$$

$$C_M = \begin{bmatrix} 1.4936e - 3 & 8.1852e - 3 & 2.7168e - 3 \end{bmatrix}$$

Define the state-variable controller matrix as

$$F^{CCF} = \begin{bmatrix} 0 & 0 & 1 \end{bmatrix}\left(\overline{A}^{CCF} - A_M \right) =$$

$$\begin{bmatrix} 0 & 0 & 1 \end{bmatrix}\left(\begin{bmatrix} 0 & 1 & 0 \\ 0 & 0 & 1 \\ .93 & -2.8564 & 2.92627 \end{bmatrix} - \begin{bmatrix} 0 & 1 & 0 \\ 0 & 0 & 1 \\ .3012 & -1.4806 & 2.167 \end{bmatrix} \right)$$

$$= \begin{bmatrix} .62887 & -1.3758 & .75927 \end{bmatrix}$$

Finally,

$$F = F^{CCF}P = \begin{bmatrix} .62887 & -1.3758 & .75927 \end{bmatrix}\begin{bmatrix} 2.8304e3 & 1.4367e3 & -1.7654e3 \\ 2.4425e3 & 1.82306e3 & -1.3752e3 \\ 2.0370e3 & 2.224e3 & -973.46 \end{bmatrix}$$

$$= \begin{bmatrix} -33.761 & 83.9036 & 42.7114 \end{bmatrix}$$

Use a computer tool to obtain the transfer function representing the dynamics of the closed-loop system defined as:

$$G_{CL}(z) = \overline{C}\left(Iz - \overline{A} + \overline{B}F \right)^{-1}\overline{B} = \frac{.1182(z - .88812)(z - .98807)}{(z - .36788)(z^2 - 1.799z + .81873)}$$

It could be seen that $G_{CL}(z)$ has the numerator of the original transfer function of the controlled plant and the denominator of the discrete-time model transfer function. This implies that the transfer function of the filter in the reference channel is to be defined as the "numerator of the discrete-time model transfer function divided by the numerator of the original transfer function of the controlled plan", i.e.

$$W(z) = \frac{2.7168e-03\,z^2 + 8.1852e-3\,z + 1.4936e-3}{.1182(z-.88812)(z-.98807)}$$

Now implement this system in the simulation environment and subject it to some tests.

The following graphs represent the implementation of the continuous-time controlled plant and the state-variable controller:

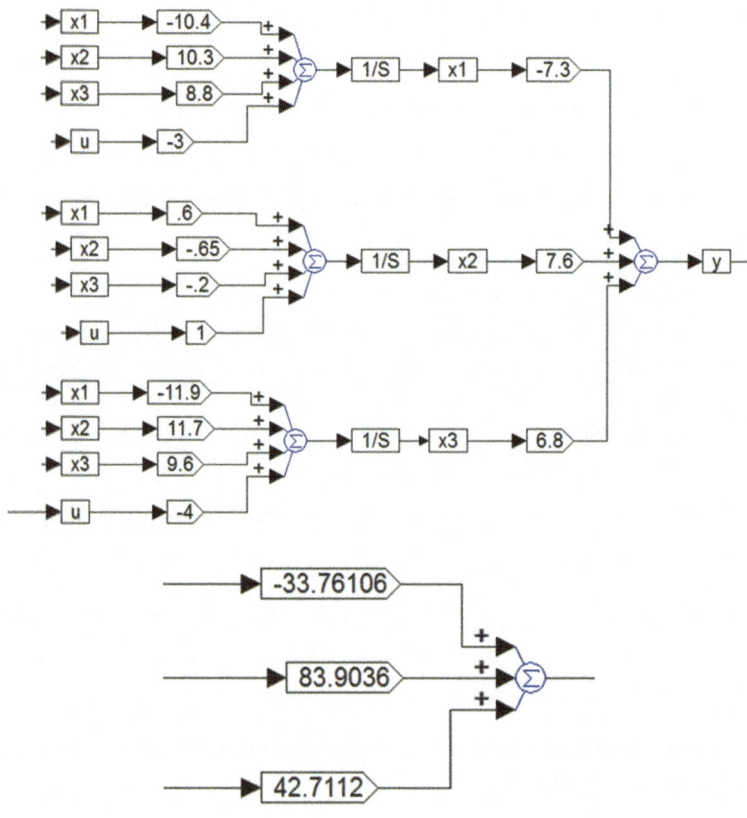

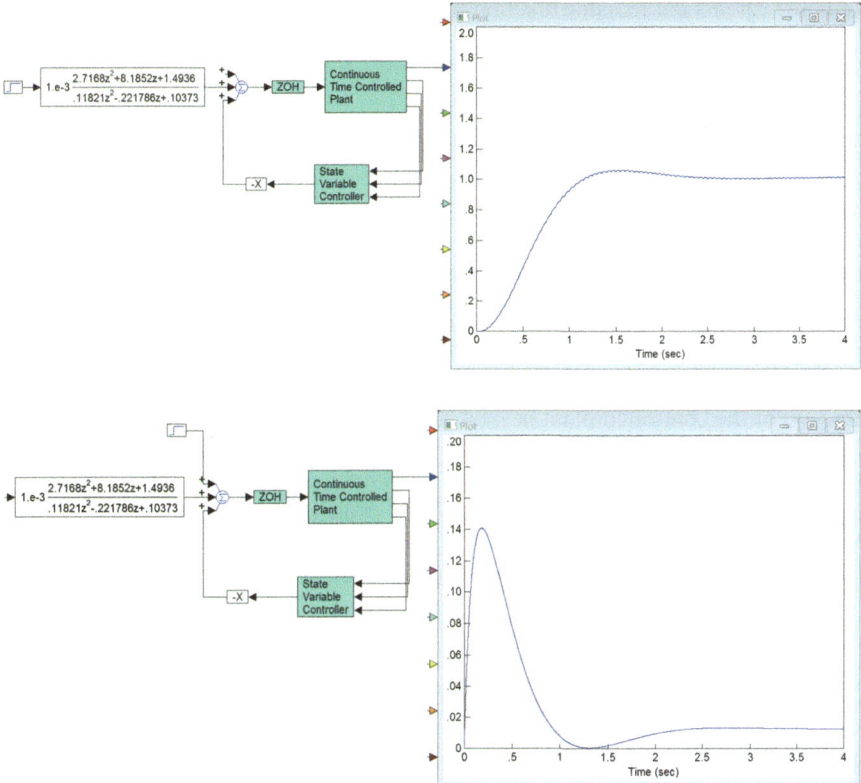

The above plots indicate that the system design specifications in terms of system dynamics and steady-state error are properly addressed.

Exercise 3.2: Problem 1

Based on the design specifications, the following s-domain model transfer function is suggested:

$$G_M(s) = \frac{26}{s^3 + 12s^2 + 40s + 26}$$

This transfer function is converted into the z-domain using the ZOH option and the time step of 0.005 s:

$$G_M(z) = \frac{5.3361e - 7\,z^2 + 2.1827e - 6\,z + 5.1784e - 7}{z^3 - 2.9408\,z^2 + 2.88256\,z - 0.94176}$$

Represent the model transfer function by a CCF:

$$X[k+1] = A_M X[k] + B_M u[k], \quad y[k] = C_M X[k]$$

where

$$A_M = \begin{bmatrix} 0 & 1 & 0 \\ 0 & 0 & 1 \\ .94176 & -2.88256 & 2.9408 \end{bmatrix}, \quad B_M = \begin{bmatrix} 0 \\ 0 \\ 1 \end{bmatrix},$$

$$C_M = [5.1784e-7 \quad 2.1027087e-6 \quad 5.33361e-7]$$

Converted into the z-domain (ZOH option and the time step of 0.005 s) the transfer function of the controlled plant is:

$$G_P(z) = \frac{0.02492\, z^2 - 0.04975\, z + 0.02483}{z^3 - 2.99\, z^2 + 2.98\, z - 0.99}$$

The CCF equivalent of this transfer function has the following matrices:

$$A_P = \begin{bmatrix} 0 & 1 & 0 \\ 0 & 0 & 1 \\ .99 & -2.98 & 2.99 \end{bmatrix}, \quad B_P = \begin{bmatrix} 0 \\ 0 \\ 1 \end{bmatrix}, \quad C_P = [.0248 \quad -.0498 \quad .0907]$$

Now, the controller matrix F can be defined as:

$$F = [0 \quad 0 \quad 0](A_P - A_M) = [.04828 \quad -.09746 \quad .04918]$$

Now, configuring the state observer. The continuous-time transfer function representing the observer dynamics is suggested as (note that the numerator of this transfer function is irrelevant):

$$G_{OBS}(s) = \frac{1}{s^3 + 83.2\, s^2 + 1280\, s + 3277}$$

The conversion of this transfer function to the z-domain (ZOH option and the time step of 0.005 s) results in the following discrete-time domain fundamental matrix of the observation process:

$$A_{OBS} = \begin{bmatrix} 0 & 1 & 0 \\ 0 & 0 & 1 \\ .6597 & -2.2934 & 2.6334 \end{bmatrix}$$

The application of the recursive formula for matrix K results in the following
K-matrix of the observer: $K = \begin{bmatrix} .3566 \\ .3796 \\ .4026 \end{bmatrix}$, and the fundamental matrix of the state
observer:

$$A_P - KC_1 = \begin{bmatrix} -.3566 & 1 & 0 \\ -.3796 & 0 & 1 \\ .5875 & -2.98 & 2.99 \end{bmatrix}$$

where matrix $C_1 = [1\ 0\ 0]$. Note that in order to obtain matrix C_1, the following filter is place in the input of the controlled plant:

$$P(z) = \frac{1}{0.02492\,z^2 - 0.04975\,z + 0.02483}$$

And finally, a special filter must be placed in the reference channel of the closed-loop system: the constant gain must be added. Since this is discrete time, there must be a delay in this block.

$$W(z) = \frac{5.3361e - 7\,z^2 + 2.1827e - 6\,z + 5.1784e - 7}{z^2}$$

The following are the simulation setup of the designed system and its responses to unit step reference and unit step disturbance signals indicating that the design requirements are successfully met.

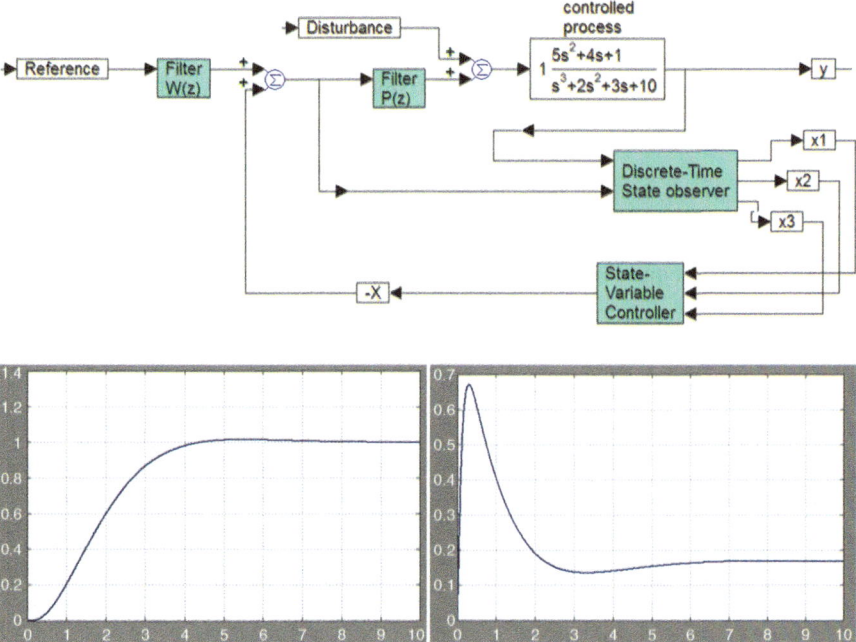

Exercise 3.2: Problem 2

First represent the controlled plant by a discrete-time transfer matrix obtained for the time step of 0.01 s and the ZOH option:

$$
G_P(z) =
\begin{bmatrix}
\dfrac{1.0048e - 2(z - .9802)}{z^2 - 1.9894z + .99} & \dfrac{4.8052e - 3}{z - .9231} \\[4mm]
\dfrac{1.4947e - 5(z + .9966)}{z^2 - 1.9876z + .99} & \dfrac{1.0095e - 2(z - .9048)}{(z - .9802)(z - .9418)}
\end{bmatrix}
$$

Now define the decoupling filter as a transfer matrix adjoint to $G_P(z)$:

$$
W(z) =
\begin{bmatrix}
\dfrac{1.0095e - 2(z - .9048)}{(z - .9802)(z - .9418)} & -\dfrac{4.8052e - 3}{z - .9231} \\[4mm]
-\dfrac{1.4947e - 5(z + .9966)}{z^2 - 1.9876z + .99} & \dfrac{1.0048e - 2(z - .9802)}{z^2 - 1.9894z + .99}
\end{bmatrix}
$$

Now it is known that the product of GP(z)W(z) is a diagonal matrix and its non-zero elements $Q_{11}(z) = Q_{22}(z) = Q(z) = \text{Det}\{G_P(z)\}$:

$$
Q(z) = \frac{N(z)}{D(z)}
$$

where polynomials N(z) and D(z) are

$$
N(z) = 1.0e - 5\left(1.0135z^5 - 48.63z^4 + 93.30153z^3 - 89.492z^2 + 42.913z - 8.2294\right)
$$

$$
D(z) = z^7 - 6.822z^6 + 19.9465z^5 - 32.4001z^4 + 31.578z^3 - 18.466z^2 + 5.9992z - .8353
$$

Introduce a 7-th order s-domain transfer function consistent with the required dynamics of decoupled channels:

$$
G_M(s) = \frac{2 \cdot 8 \cdot 9 \cdot 10 \cdot 11 \cdot 12 \cdot 13}{(s + 2)(s + 8)(s + 9)(s + 10)(s + 11)(s + 12)(s + 13)}
$$

and its z-domain equivalent (for ZOH option and time step of 0.01 s):

$$
G_M(z) = \frac{N_M(z)}{D_M(z)}
$$

We chose not to offer explicit expressions for polynomials $N_M(z)$ and $D_M(z)$ that could obtained via a software tool, however, the components of the control systems stabilizing the decoupled channels, identical for both channels, are defined as:

feedback controller:

$$H(z) = \frac{D_M(z) - D(z)}{N(z)}$$

filter in the reference channel:

$$P(z) = \frac{N_M(z)}{D(z)}$$

The implementation of the obtained results and simulation based analysis is clarified by the figure below that features a continuous-time two-input-two-output controlled process driven by discrete-time circuitry.

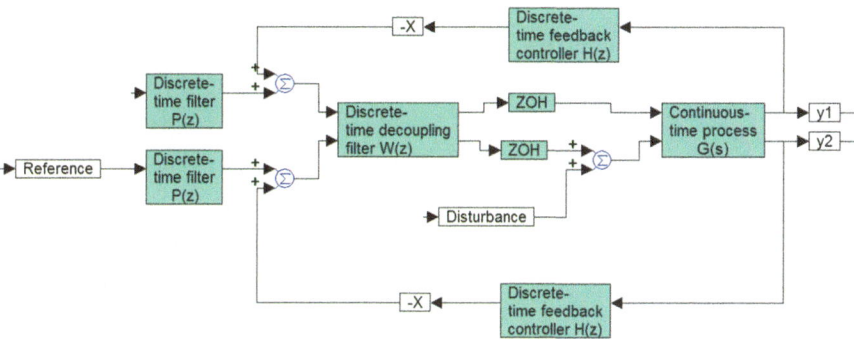

Exercise 3.2: Problem 3

Obtain the matrix–vector description of the controlled plant:

$$\dot{X} = A_P X + B_P U, \quad Y = C_P X$$

where

$$A_P = \begin{bmatrix} 4 & 3 & 8 \\ 6 & -5 & -2 \\ -1 & 7 & 9 \end{bmatrix}, \quad B_P = \begin{bmatrix} -3 & 1 & -1 \\ 1 & 2 & 1 \\ -4 & 1 & -5 \end{bmatrix}, \quad C_P = \begin{bmatrix} -3 & 7 & 4 \\ -1 & 1 & -2 \\ 1 & 3 & -1 \end{bmatrix}$$

The discrete-time equivalent of the above description for ZOH option and time step $\Delta t = 0.01$ s is found as follows:

$$A = \begin{bmatrix} 1.0413 & .03273 & .0851 \\ .0598 & .9515 & -.0179 \\ -.0085 & .0713 & 1.0931 \end{bmatrix}, \quad B = \begin{bmatrix} -.0321 & .0109 & -.0121 \\ .0092 & .0197 & .0099 \\ -.0413 & .0111 & -.0519 \end{bmatrix},$$

$$C = C_P = \begin{bmatrix} -3 & 7 & 4 \\ -1 & 1 & -2 \\ 1 & 3 & -1 \end{bmatrix}$$

Next, each channel's desired transfer function can be represented by a first order s-domain transfer function based on the required settling time:

$$G_1(s) = \frac{\frac{4}{T_{set}}}{s + \frac{4}{T_{set}}} = \frac{1.33}{s + 1.33} \quad G_2(s) = \frac{\frac{4}{T_{set}}}{s + \frac{4}{T_{set}}} = \frac{0.667}{s + 0.667} \quad G_3(s) = \frac{\frac{4}{T_{set}}}{s + \frac{4}{T_{set}}} = \frac{0.5}{s + 0.5}$$

Next, these transfer functions are converted into z-domain transfer functions (ZOH option and $\Delta t = 0.01$ s):

$$G_1(z) = \frac{0.01324}{z - 0.9868} \quad G_2(z) = \frac{0.0066}{z - 0.9934} \quad G_3(z) = \frac{0.005}{z - 0.995}$$

These transfer functions are used to defined the following matrices representing the desired closed-loop three-input-three-output decoupled system:

$$Q = \begin{bmatrix} 0.9868 & 0 & 0 \\ 0 & 0.9934 & 0 \\ 0 & 0 & 0.995 \end{bmatrix}$$

$$P = \begin{bmatrix} 0.01324 & 0 & 0 \\ 0 & 0.0066 & 0 \\ 0 & 0 & 0.005 \end{bmatrix}$$

Finally, matrix W of the filter is calculated as:

$$W = (C \cdot B)^{-1} \cdot P$$

$$W = \begin{bmatrix} 0.0818 & 0.0805 & -0.0644 \\ 0.0389 & -0.0169 & 0.0445 \\ -0.0764 & -0.0284 & 0.0682 \end{bmatrix}$$

And matrix F of the controller is calculated as:

$$F = (C \cdot B)^{-1} \cdot (C \cdot A - Q \cdot C)$$

$$F = \begin{bmatrix} -1.3102 & -0.8201 & -2.5295 \\ 2.6645 & -1.1310 & 0.3051 \\ 1.8059 & -1.0389 & 0.1228 \end{bmatrix}$$

The calculated control system was simulated in Simulink with the following model:

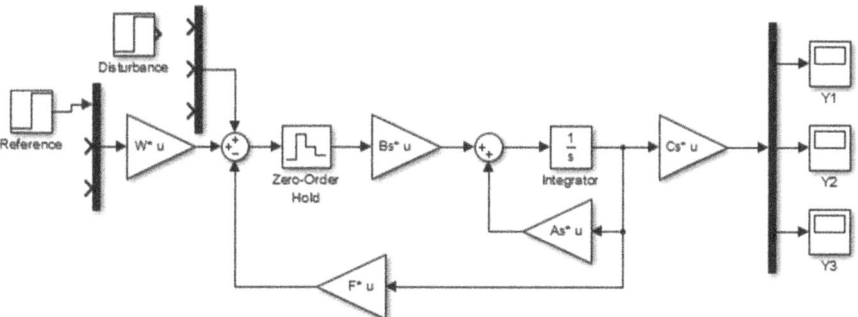

The decoupling effect and compliance with the settling time requirements the following simulation results are shown below.

1. Unit step reference signal applied to input #1 results in the response of output #1 reaching the steady state value of 1.0 with the settling time of approximately 3 s. System responses observed in channels #2 and #3 have very low magnitudes.

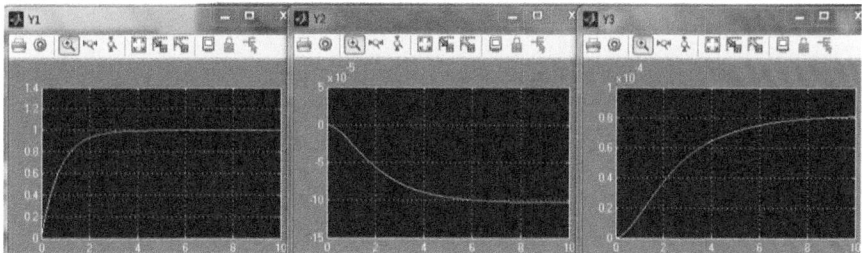

Similar results could be observed when unit step reference signal was applied to the inputs #2 and #3, see below.

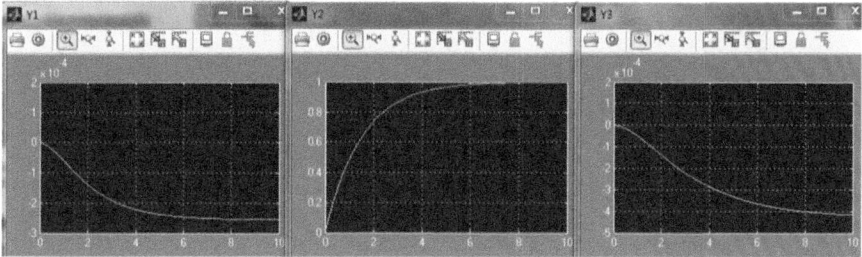

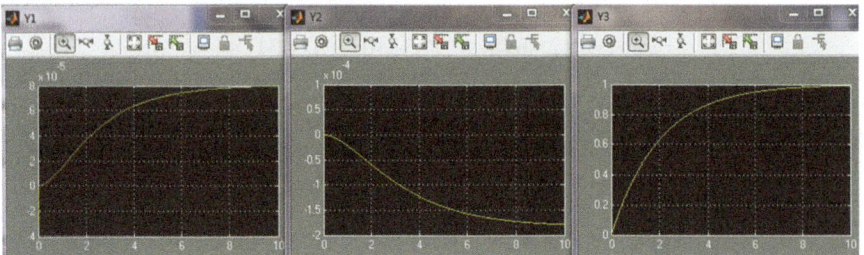

Exercise 3.3: Problem 1

The output of the plant alone was simulated first with this input to demonstrate the need for a controller in the system. The system does not have the desired settling time or overshoot of the step response, so the system must be controlled to achieve such an output.

Next, a second order model was created in the s-domain to meet the design specifications. The dominant pole is chosen to be a first order pole equal to 4/Tset such that this pole will control the settling time. The dominant pole is -0.667 and the non-dominant pole is chosen to be -10 simply to address the possible error requirement.

$$G_M(s) = \frac{6.67}{s^2 + 10.67s + 6.67}$$

The step response for this model is plotted below. As seen in the step response plot, the settling time is almost exactly 6 s and there is 0% overshoot, so this model meets the design specifications.

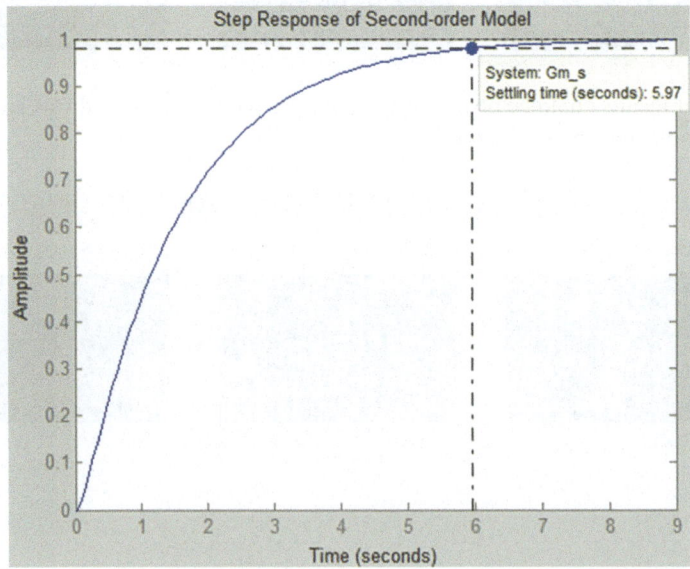

Then, for discrete-time control, the model is converted into the z-domain using the zero-order-hold option and the time step provided of 0.05 s. The resulting discrete model transfer function is

$$G_M(z) = \frac{0.007024z + 0.005882}{z^2 - 1.574z + 0.5866}$$

And its step response is below. It still meets all of its design specifications, as is to be expected.

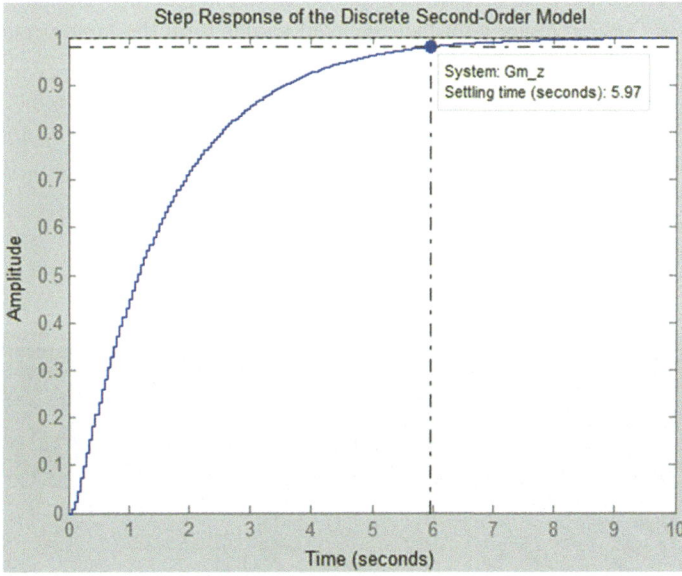

The closed-loop transfer function can be defined as

$$G_{CL}(z) = \frac{N_F(z)}{D_P(z) + N_C(z)} = G_M(z) = \frac{N_M(z)}{D_M(z)}$$

Following the approach outlined in Sect. 3.3

$$N_C(z) = \alpha_1 z + \alpha_0 \text{ and } D_C(z) = b_1 z + b_0$$

and

$$U_{NEW}(z) \cdot (b_1 z + b_0) + Y_{NEW}(z) \cdot (h_1 z + h_0) = Y(z)$$

where

$$U_{NEW}(z) = \left(\frac{U(z)}{z^2 - 1.574z + 0.5866} \right)$$

$$Y_{NEW}(z) = \left(\frac{Y(z)}{z^2 - 1.574z + 0.5866} \right)$$

To estimate parameters b_1, b_0, h_1 and h_0 we use the RLSM that is performed inside a function block in Simulink. The simulation model obtains parameter values from the RLSM block and used them to tune the controller in real time. Signals Y and U subjected to filtering along with their delayed values form the input vector of the RLSM block.

The simulation ran for 10 s, and the results are:

b_1	0.002391
b_0	0.01061
h_1	0.07248
h_0	−0.07808

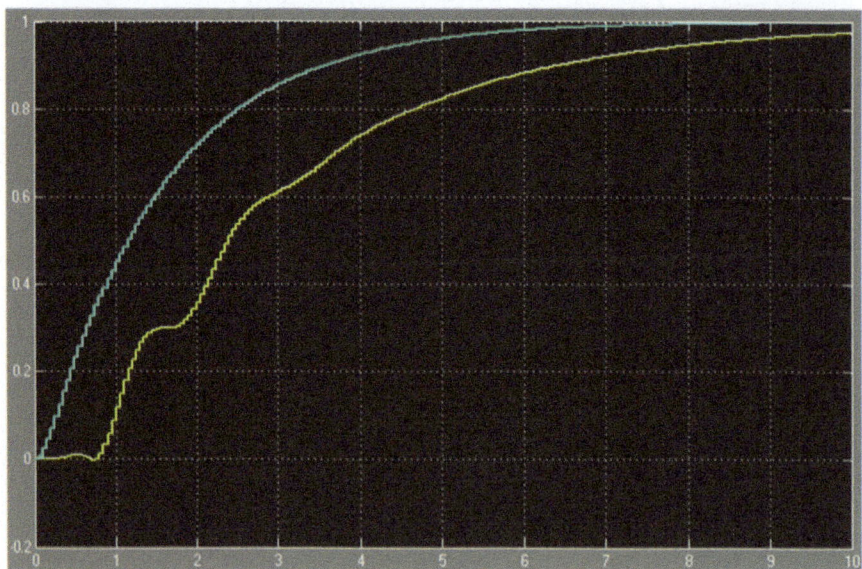

The output of the controlled system and the output of the model

It could be seen that the system output (yellow) converges to the output of the chosen model (blue) representing the design requirements.

The entire Simulink setup is shown below.

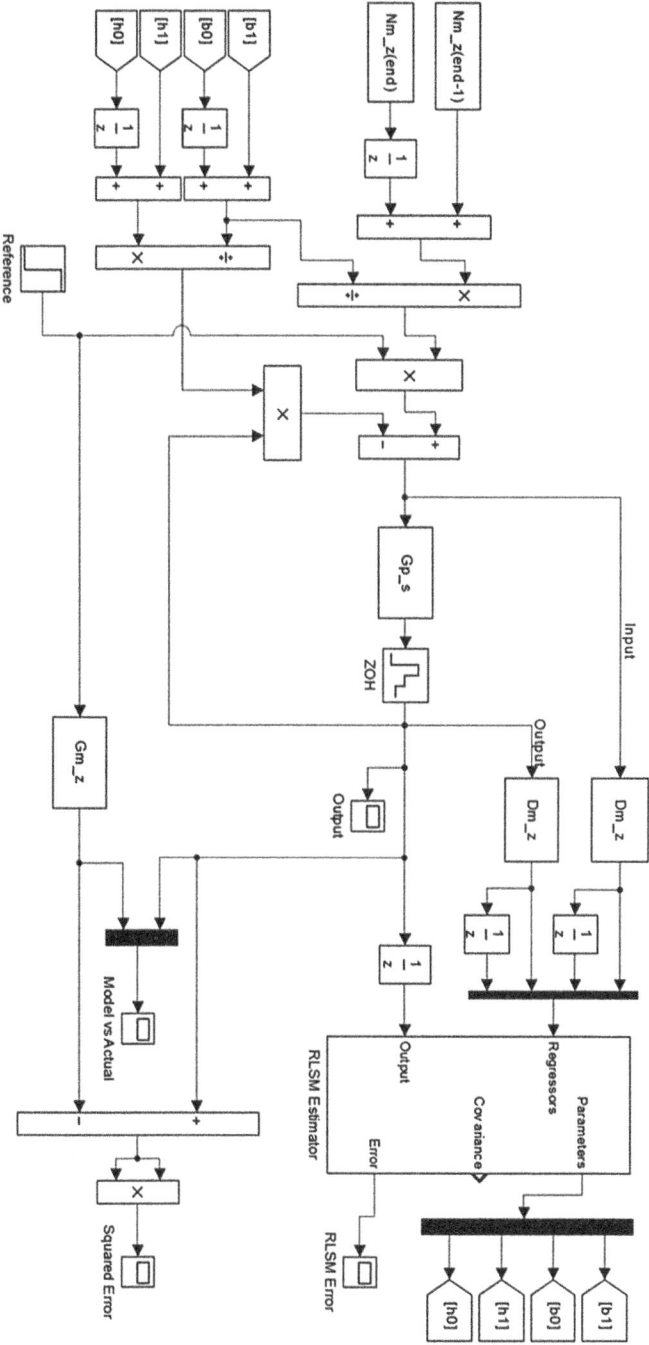

Bibliography

Skormin, V., Introduction to Automatic Control, Volume I, Linus Publications, Inc, ISBN 1-60797-090-2

Skormin, V., Introduction to Automatic Control, Volume II, Linus Publications, Inc, ISBN 1-60797-162-3

Ogata, K. Discrete-time control systems, Prentice-Hall, ISBN0-13-216102-8

Astrom, K.J. and Wittenmark, B., Adaptive Control, Addison-Wesley Publishing Company, ISBN 0-201-09720-6

MATLAB: https://www.mathworks.com/products/matlab-home/

Simulink: http://www.mathworks.com/products/simulink/

VISSIM: http://www.vissim.com/

Chapter 4
Methods and Models of Optimization

Most engineering problems, including planning, control and design, have more than one solution. The theory of optimization provides a mathematical basis for establishing the acceptability conditions that outline the class of acceptable solutions, for the definition of the criterion that provides the measure of goodness of every individual solution, and the optimization procedure (algorithm) that results in finding the optimal solution, i.e. the solution maximizing the value of the goodness criterion. These three components, the *class of acceptable solutions*, the *criterion of goodness*, and the *optimization procedure* are to be present in any definition of the optimization problem.

The solution vector of the optimization problem is a set of particular numerical values of some optimization variables, $X = [x_1, x_2, \ldots, x_n]^T$ that represent the nature of the problem. For example, in a resource distribution problem when some material, monetary, or energy resources have to be distributed between n consumers, vector X represents the amounts of these resources designated to each consumer.

The class of acceptable solutions is typically defined as a set of conditions, equations and/or inequalities that the solution vector must satisfy. Thus in the resource distribution problem these conditions include the requirements that the amounts of the resource designated to individual consumers cannot be negative, i.e.

$$x_i \geq 0, \ i = 1, 2, \ldots, n$$

that the sum of the amounts of this resource designated to consumers shall not exceed the total amount available ($i = 1, 2, \ldots$ is the consumer index), i.e.

$$\sum_{i=1}^{n} x_i \leq P^{TOT}$$

© Springer International Publishing Switzerland 2016
V.A. Skormin, *Introduction to Process Control*, Springer Texts
in Business and Economics, DOI 10.1007/978-3-319-42258-9_4

that the amounts of the resources provided to some of the consumers are not negotiable (k is the consumer index), i.e.

$$x_k = P^K, \ k = k_1, \ k_2, \ \ldots$$

or shall have allowable minimal and maximal values, i.e.

$$P_{MIN}^K \leq x_k \leq P_{MAX}^K, \ k = k_1, \ k_2, \ \ldots$$

It is said that the conditions outlining the class of acceptable solutions reflect the feasibility requirements and the specifics of the problem, and form a special region in the solution space X.

The optimization criterion is always a scalar function defined in the solution space

$$Q(X) = Q(x_1, x_2, \ \ldots, \ x_n)$$

that represents the degree of the consistence of any solution vector X to the general goal of the engineering task. For example, the resource distribution problem may reflect the goal of maximizing the resource utilization, and intuitively its solution would provide maximum allowable amounts of the resource to the consumers having the highest coefficients of its utilization, α_i, $i = 1,2,\ldots,n$. It could be seen that in this case the criterion could be defined as

$$Q(X) = \sum_{i=1}^{n} \alpha_i x_i$$

In the situation when the goal of the resource distribution problem is to minimize the total cost of transporting the resource to the consumers, and intuitively the most remote consumers are expected to receive the least amounts of the resource within the allowable limits, the criterion could be visualized as

$$Q(X) = \sum_{i=1}^{n} \beta_i x_i$$

where β_i, $i = 1,2,\ldots$, n are the transportation costs per unit of the resource for particular consumers. It is common to refer to the function $Q(X)$ as **criterion**, or **objective function**, or a **loss function**, that highlights various aspects of the nature of the optimization problem.

Finally, the optimization procedure must result in the rule that would facilitate the detection of such a point, X^{OPT}, in the region of acceptable solutions in the space X where criterion $Q(X)$ has its minimum (maximum) value

$$Q^{OPT} = Q(X^{OPT})$$

One should realize that the search for the maximum of criterion $Q_1(X)$ is equivalent to the search of the minimum of the criterion $Q_2(X) = -Q_1(X)$ and vice versa, therefore we will always refer to the task of *optimization* as the task of *minimization*.

Recall the approach to minimization presented as a part of undergraduate calculus. It suggests that if a minimum of some scalar function

$$Q(X) = Q(x_1, x_2, \ldots, x_n),$$

i.e. $X^* = [x_1^*, x_2^*, \ldots, x_n^*]$, could be found as the solution of the system of n equations,

$$\frac{\partial}{\partial x_1} Q(x_1, x_2, \ldots, x_n) = f_1(x_1, x_2, \ldots, x_n) = 0$$

$$\frac{\partial}{\partial x_2} Q(x_1, x_2, \ldots, x_n) = f_2(x_1, x_2, \ldots, x_n) = 0$$

$$\cdots\cdots\cdots\cdots\cdots\cdots\cdots\cdots\cdots\cdots\cdots\cdots\cdots\cdots\cdots\cdots$$

$$\frac{\partial}{\partial x_n} Q(x_1, x_2, \ldots, x_n) = f_n(x_1, x_2, \ldots, x_n) = 0$$

While this suggestion is consistent with the rigors of a community college, from the engineering point of view it is quite unrealistic because of the following reasons,

1. Function $Q(x_1, x_2, \ldots, x_n)$ is typically so complex that its derivatives

$$\frac{\partial}{\partial x_i} Q(x_1, x_2, \ldots, x_n), \ i = 1, 2, \ldots, n$$

   are very difficult and sometimes impossible to define analytically
2. Derivatives $\frac{\partial}{\partial x_i} Q(x_1, x_2, \ldots, x_n), \ i = 1, 2, \ldots, n$ are nonlinear functions of x_1, $x_2, \ldots, x_n$, and the system of equations shown above may have multiple solutions, may have no solution, and in any case, cannot be solved analytically.
3. The entire definition of the function minimization task does not address the existence of constraints.
4. The function to be minimized may not have any analytical definition, but for any combination of numerical values of its arguments its value could be defined numerically, for example by conducting an experiment.

In most real life situations, the optimization task could be performed only numerically and be compared with the navigation through a very complex terrain to the highest (in the maximization case) existing peak while avoiding the obstacles and low peaks. The task is aggravated by the fact that the terrain is multidimensional and the obstacles could be detected only by direct contact. Figure 4.1 below drawn by my cartoonist friend Joseph Kogan depicts the task of optimization based on my comments.

Unsurprisingly, the optimization became an engineering tool only due to the proliferation of modern computers. We will present several common models,

Fig. 4.1 Myth vs. reality of optimization

methods and applications of optimization that should be included in the toolbox of a modern engineer. These techniques will include linear programming, numerical techniques of nonlinear programming (gradient, random and direct search), genetic optimization, and dynamic programming. We do not expect a modern engineer to develop optimization techniques, this is the mathematicians' domain, however a good engineer shall be able to:

- recognize a situation lending itself to an optimization task
- formulate the optimization problem, i.e. define its variables, criterion and constraints
- recognize the resultant problem as one of the typical optimization problems
- find and apply a suitable optimization tool (perhaps available in MATLAB)

4.1   Linear Programming

Linear programming is an optimization technique suitable for the situations when the set of conditions, outlining the region of acceptable solutions, and the goodness criterion are linear functions defined in the solution space.

In a linear programming problem, the region of acceptable solutions is defined by the set of equalities and inequalities as follows:

$$\sum_{i=1}^{n} a_{ij}x_i = b_j \text{ and } \sum_{i=1}^{n} a_{iK}x_i \leq b_K$$

where x_i, $i = 1,2,..,\ n$ are optimization variables that constitute the solution space, $j = 1,2,\ldots,\ L$ is the equality index, and $k = 1,2,\ldots,\ M$ is the inequality index. Note that the number of equalities must be less than the dimension of the solution space otherwise the region of the acceptable solutions will include only one point (when $n = L$), or could be empty (when $L > n$). One should understand that inequalities can always be redefined as the standard "greater or equal" type, indeed inequality of "less or equal" type, i.e. $\sum_{i=1}^{n} a_{iK} x_i \leq b_K$ could be easily converted into the "greater or equal" type by changing signs: $-\sum_{i=1}^{n} a_{iK} x_i \geq -b_K$, consequently only the "greater or equal" type inequalities will be considered. Note that the class of acceptable solutions could be empty even when $n > L$: the inequalities and equalities could be mutually contradictive.

The criterion of a linear optimization problem is defined by a linear function,

$$Q(x_1, x_2, \ldots, x_n) = \sum_{i=1}^{n} c_i x_i$$

that has to be minimized,

$$Q(x_1, x_2, \ldots, x_n) = \sum_{i=1}^{n} c_i x_i \rightarrow \min$$

or

$$-Q(x_1, x_2, \ldots, x_n) = -\sum_{i=1}^{n} c_i x_i \rightarrow \min$$

if the original criterion $Q(X)$ had to be maximized.

Example 4.1 Consider one of the typical problems of linear programming, the task distribution problem. There are 5 reactors operating at a chemical plant and producing the same product. Due to capacity, design specifics and the technical status, the reactors have different efficiency expressed by the extraction coefficients, α_j, $j = 1,2,3,4,5$. The capacities of these reactors, q_j, $j = 1,2,3,4,5$, are also different

reactor, j	1	2	3	4	5
coefficient α_j	0.81	0.76	0.63	0.71	0.68
capacity q_j (units)	150	200	175	120	96

The chemical plant is required to process a certain amount of raw material, say $P = 500$ units that should be rationally distributed between the reactors in the sense that the overall extraction coefficient will be maximized. It could be seen that the solution space of this problem comprises of 5 variables, x_1–x_5, representing the amount of raw material loaded in respective reactors. The constraints of this problem must address the following requirements:

Amount of raw material loaded in the j-th reactor must be non-negative: $x_j \geq 0$, $j = 1,2,\ldots,5$

The total amount of raw material to be loaded in reactors is defined: $\sum_{j=1}^{5} x_j = P$

The amount of raw material loaded in a particular reactor cannot exceed the capacity of this reactor: $x_j \leq q_j$, $j = 1,2,\ldots,5$

The criterion of this problem could be defined as $\sum\limits_{j=1}^{5} a_j x_j \to$ max   or

$-\sum\limits_{j=1}^{5} a_j x_j \to$ min

The mathematical formulation of this problem is

$$-0.81x_1 - 0.76x_2 - 0.63x_3 - 0.71x_4 - 0.68x_5 \to \text{min}$$

subject to conditions

$$x_1 \geq 0, x_2 \geq 0, x_3 \geq 0, x_4 \geq 0, x_5 \geq 0$$
$$-x_1 \geq -150, \; -x_2 \geq -200, \; -x_3 \geq -175, \; -x_4 \geq -120, \; -x_5 \geq -96$$
$$x_1 + x_2 + x_3 + x_4 + x_5 = 500$$

One can realize that this problem has an infinite number of alternative solutions providing that the total amount of raw material P is less than the total capacity of the reactors, thus creating the opportunity for the optimization. In the case when the total amount of raw material P is equal to the total capacity of the reactors, only one solution exists and the optimization is impossible. Finally, in the case when the total amount of raw material P is greater than the total capacity of the reactors, the problem does not have any solution.

It could be also realized that the optimal solution procedure for this problem is quite trivial:

Step 1.   The first reactor, having the highest efficiency coefficient, should be loaded to full capacity ($x_1^{OPT} = 150$, 350 units still is to be distributed), then

Step 2.   The second most efficient reactor must be loaded to full capacity ($x_2^{OPT} = 200$, 150 units is to be distributed), then

Step 3.   The third most efficient reactor must be loaded to full capacity ($x_4^{OPT} = 120$, 30 units is to be distributed), then

Step 4.   The fourth most efficient reactor must be loaded with the remaining amount of raw material ($x_5^{OPT} = 30$ units, zero units to be distributed), then $x_3^{OPT} = 0$.

It should be noted that most linear programming problems do not allow for such a simple solution procedure.

Example 4.2 The transportation problem. A product stored at 3 warehouses must be distributed between 5 consumers in such a fashion that the total cost of transporting the product is minimized.

The solution space of this problem is formed by $3 \times 5 = 15$ variables, x_{jk}, $j = 1,2,3$, $k = 1,2,3,4,5$, representing the amount of the product delivered from the j-th warehouse to the k-th consumer. Introduce the matrix of transportation costs, c_{jk}, $j = 1,2,3$, $k = 1,2,3,4,5$, representing the cost of transportation of one unit of the product from the j-th warehouse to the k-th consumer. Introduce quantities P_j, $j = 1,2,3$, representing the amount of the product at j-th warehouse, and quantities W_k, $k = 1,2,3,4,5$, representing the amount of the product requested by k-th consumer. Then the mathematical formulation of the problem is

$$\sum_{k=1}^{5} \sum_{j=1}^{3} c_{jk} x_{jk} \rightarrow \min$$

subject to the following conditions

a) non-negativity, $x_{jk} \geq 0$, $j = 1,2,3$, $k = 1,2,3,4,5$

b) amount of the product available at each warehouse, $\sum_{k=1}^{5} x_{jk} \leq P_j$, $j = 1,2,3$

c) amount of product delivered to each consumer, $\sum_{j=1}^{3} x_{jk} = W_k$, $k = 1,2,3,4,5$

One can realize that the solution of this problem exists if $\sum_{k=1}^{5} W_k \leq \sum_{j=1}^{3} P_j$, however it cannot be obtained without a computationally intensive and rigorously justified algorithm. It also should be noted that typical solutions of linear programming problems comprise non-negative variables and therefore the non-negativity of the solution is assured not by special constraints but by the solution procedure itself.

Example 4.3 The mixing problem. Preparing the right raw material is one of the conditions for obtaining a high quality end product in chemical or metallurgical manufacturing. Assume that the raw material is characterized by percentages of four ingredients: $A_1\%$, $A_2\%$, $A_3\%$, and $A_4\%$. The raw material is prepared by mixing six components in the amounts (in tons) $x_1, x_2, \ldots, x_6$. Each component contains all four ingredients, but the concentrations are all different, for example a_{jk} (%) is the concentration of the ingredient #j ($j = 1,2,3,4$) in the component #k ($k = 1,2,3,4,5,6$). The cost of each component is given: c_k ($/ton), ($k = 1,2,3,4,5,6$). Also given are the required total amount of the raw material, P (tons) and the available amounts of the individual components, q_k (tons), ($k = 1,2,3,4,5,6$). It is required to prepare the least expensive mixture.

The problem definition is as follows:
Minimize the cost of the mixture:

$$\sum_{k=1}^{6} c_k x_k \rightarrow \min$$

Subject to constraints on

the total amount of the raw material $\sum\limits_{k=1}^{6} x_k = P$

percentages of four ingredients $(j = 1,2,3,4)$ $\sum\limits_{k=1}^{6} a_{jk} x_k = A_j \cdot P$

available amounts of individual components, $(k = 1,2,3,4,5,6)$ $x_k \leq q_k$

Again, the optimal problem solution, if it exists, could be obtained via some numerically extensive procedure.

Let us consider such a procedure.

4.1.1 Geometrical Interpretation of Linear Programming

Geometrical interpretation of linear programming is crucial for the understanding of the computational nature of its algorithm. Geometrical interpretation works best for the two-dimensional solution space and the inequality-type constraints.

Consider a straight line in two-dimensional space defined by the equation $a_1 x_1 + a_2 x_2 = b$ like the one below in Fig. 4.2.

It is known that any point on this line, for example $[x_1^1, x_2^1]$ satisfies this equation, i.e. $a_1 x_1^1 + a_2 x_2^1 = b$. It also known that any point above this line, such as $[x_1^2, x_2^2]$, results in $a_1 x_1^2 + a_1 x_2^2 > b$, and any point below this line, $[x_1^3, x_2^3]$, results in $a_1 x_1^3 + a_2 x_2^3 < b$. Consequently, any condition $a_1 x_1 + a_2 x_2 \leq b$ (or $-a_1 x_1 - a_2 x_2 \geq -b$) outlining the class of acceptable solutions indicates that the acceptable solutions must be located on or below the appropriate straight line. At the same time, any condition $a_1 x_1 + a_2 x_2 \geq b$ (or $-a_1 x_1 - a_2 x_2 \leq -b$) indicates that acceptable solutions must be located on or above the appropriate straight line. One can visualize a domain of acceptable solutions defined by inequality-type conditions as the part of the plane that simultaneously complies with all inequality-type conditions (highlighted below in Fig. 4.3):

Fig. 4.2 How linear constraints work

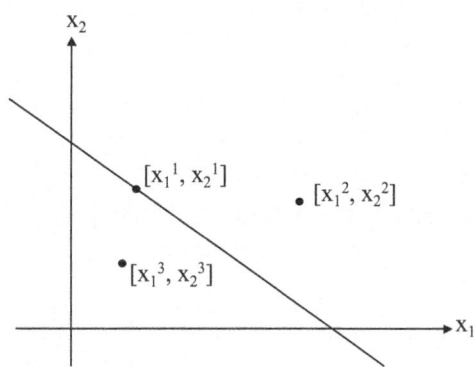

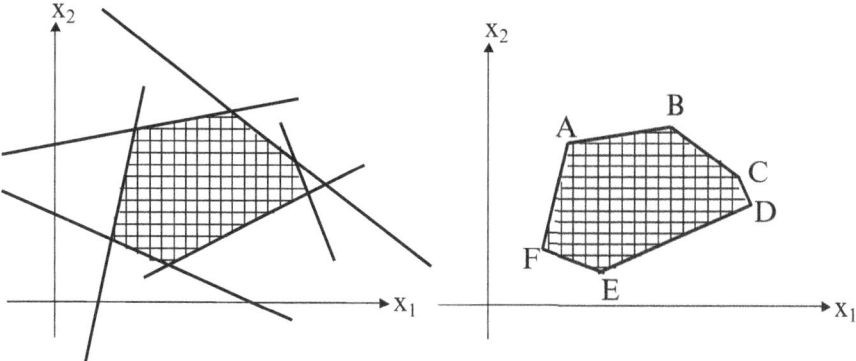

Fig. 4.3 Combination of linear constraints and domain of acceptable solutions

Fig. 4.4 Solution point and
criterion value

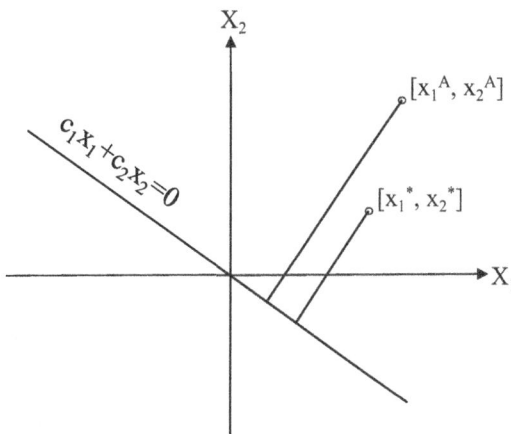

Now consider a straight line $c_1x_1 + c_2x_2 = 0$ and two points, $[x_1^A, x_2^A]$ and $[x_1^*, x_2^*]$, located in the two-dimensional space. Note that the distance between the straight line and point $[x_1^A, x_2^A]$ is greater than the distance between this line and point $[x_1^*, x_2^*]$. This results in the following result that could be easily verified by a numerical example: $c_1x_1^A + c_2x_2^A > c_1x_1^* + c_2x_2^*$.

Consider the combination of the domain of acceptable solutions bounded by contour ABCDEF and the straight line $c_1x_1 + c_2x_2 = 0$ representing the criterion of a minimization problem seen below in Fig. 4.4. Note that the domain of acceptable solutions bounded by contour ABCDEF, generally speaking, forms a *convex polyhedron* in the n-dimensional space, and its individual vertices (corner points), i.e. A, B, C, ..., are known as *basic acceptable solutions* of the linear programming problem.

It could be concluded that the solution of the problem $[x_1^{OPT}, x_2^{OPT}]$ minimizing the criterion $Q(x_1, x_2) = c_1x_1 + c_2x_2$ is located in the point that belongs to the domain of acceptable solutions and has the shortest distance from the straight line

Fig. 4.5 Graphical
interpretation of a linear
programming problem

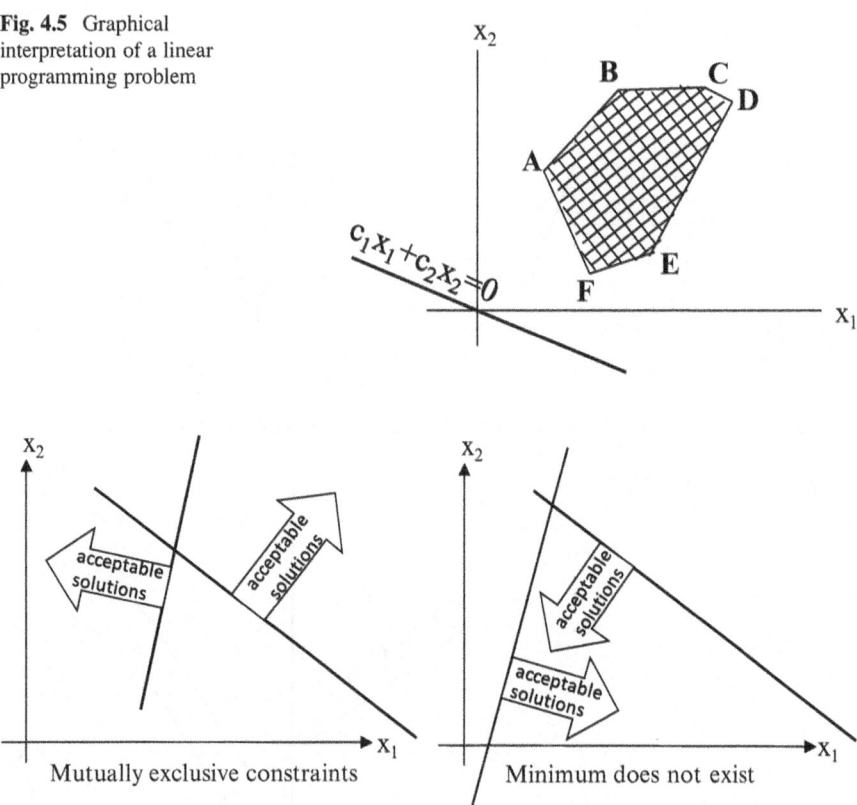

Fig. 4.6 Situations when the solution does not exist

$c_1x_1 + c_2x_2 = 0$. It could be seen that in the above Fig. 4.5 this point is F. Should the solution maximizing the criterion $Q(x_1,x_2) = c_1x_1 + c_2x_2$ be sought, it will be found in the point D that belongs to the domain of acceptable solutions and has the largest distance from the straight line $c_1x_1 + c_2x_2 = 0$.

Now consider the specifics of the linear programming problem preventing us from obtaining its optimal solution. The first condition is caused by the situation where at least two constraints are mutually exclusive, in this case even acceptable solutions do not exist. In the second case, the domain of acceptable solutions is not empty but unbounded, thus the solution minimizing the criterion does not exist. Both cases are shown in Fig. 4.6. Finally, Fig. 4.7 represents the situation where no unique optimal solution minimizing the criterion exists: the straight line representing the criterion is parallel to the side AB of the domain of acceptable solutions.

So far our discussion addressed only the inequality-type constraints. Imagine that a linear programming problem contains **k** equality-type constraints, **m** inequality-type constraints and has **n** solution variables where $n > k$. Assume that the problem is formulated as follows:

Fig. 4.7 No unique
minimum exists

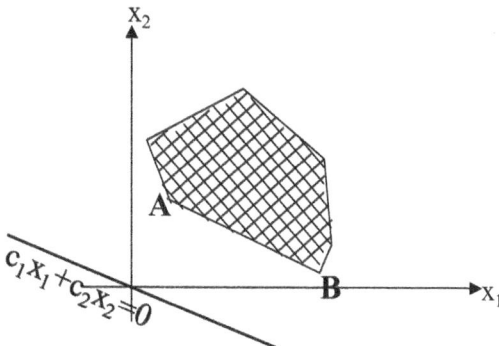

minimize $\displaystyle\sum_{i=1}^{n} c_i x_i$

subject to constraints $\displaystyle\sum_{i=1}^{n} p_{ij} x_i = q_j$, $j = 1, 2, 3, \ldots, k$

and $\displaystyle\sum_{i=1}^{n} a_{ij} x_i \leq b_j$, $j = 1, 2, 3, \ldots, m$

Note that condition $n > k$ creates the situation when k variables could be assigned arbitrary values and removed from the list of solution variables. Since our goal is the minimization of the criterion $\displaystyle\sum_{i=1}^{n} c_i x_i$ we shall assign zero values preferably to those variables that have largest values of the corresponding coefficients c_i. This is done by sequential application of a special computational operation known in linear algebra as *pivoting*. Indeed, after k pivoting steps the problem will be reduced to the following definition:

minimize $\displaystyle\sum_{i=1}^{n-k} \bar{c}_i x_i$

subject to constraints $\displaystyle\sum_{i=1}^{n-k} \bar{a}_{ij} x_i \leq \bar{b}_j$, $j = 1, 2, 3, \ldots, m$

where $\bar{a}_{ij}$, $\bar{b}_j$, $\bar{c}_j$, $i = 1, 2, \ldots, n - k, j = 1, 2, 3, \ldots m$ are problem parameters modified by pivoting steps.

In summary, a linear programming procedure intended for solution of a minimization problem with n variables, k equality-type and m inequality-type constraints ($n > k$), could be formulated as follows:

Step 1.   Reduction of the dimension of the solution space by the elimination of k strategically chosen variables and setting their values in the optimal solution to zero

Step 2.   Finding basic acceptable solutions of the problem by solving possible combinations of $n-k$ out of m equations $\sum_{i=1}^{n-k} \overline{a}_{ij} x_i = \overline{b}_j$, $j = 1, 2, 3, \ldots, m$

Step 3.   Finding the optimal solution of the problem as the basic acceptable solution that $\leq$

Note that there are many highly efficient software tools that could be recommended for the solution of a linear programming problem. (For example see http://www.onlinecalculatorfree.org/linear-programming-solver.html).

Example 4.4 Solving a simple linear programming problem given below:

$$\text{Minimize } Q(X) = 3x_1 + 10x_2 + 5x_3 + 2x_4$$

subject to conditions

$$x_1 + x_2 + x_3 + x_4 \leq 125$$
$$x_2 - 8x_3 + x_4 \leq 12$$
$$-x_1 + 2x_2 - 3x_3 + x_4 \leq 24$$
$$x_1 + x_2 = 36$$
$$2x_1 - 5x_2 + 8x_3 + 4x_4 = 16$$

The optimal solution (as per tool http://www.onlinecalculatorfree.org/linear-programming-solver.html):

$$Q^{OPT} = 164; \ x_1 = 28, \ x_2 = 8, \ x_3 = 0, \ x_4 = 0$$

Example 4.5 A resource distribution problem. A product available from three suppliers is to be provided to four consumers. The amounts of the product requested by individual consumers are respectively: 150, 230, 80 and 290 (units). The amounts of the product available at each supplier are: 300, 270 and 275 units. The transportation costs of the product from each supplier to each consumer in $ per unit are listed in the table below:

	Consumer #1	Consumer #2	Consumer #3	Consumer #4
Supplier #1	25	16	33	48
Supplier #2	45	15	36	11
Supplier #3	21	31	40	52

It is required to minimize the overall transportation cost while satisfying the consumers' demands and not to exceed suppliers' capabilities. The following problem definition is self-explanatory and at the same time is fully consistent with the data format of the tool offered at

http://www.onlinecalculatorfree.org/linear-programming-solver.html

Maximize $p = -25x_{11} - 16x_{12} - 33x_{13} - 48x_{14} - 45x_{21} - 15x_{22} - 36x_{23} - 11x_{24}$
$- 21x_{31} - 31x_{32} - 40x_{33} - 52x_{34}$

subject to

$$x_{11} + x_{12} + x_{13} + x_{14} <= 300$$

$$x_{21} + x_{22} + x_{23} + x_{24} <= 270$$

$$x_{31} + x_{32} + x_{33} + x_{34} <= 275$$

$$x_{11} + x_{21} + x_{31} + x_{41} = 150$$

$$x_{12} + x_{22} + x_{32} + x_{42} = 230$$

$$x_{13} + x_{23} + x_{33} + x_{43} = 80$$

$$x_{14} + x_{24} + x_{34} + x_{44} = 290$$

The Optimal Solution: $p = -13{,}550$; $x_{11} = 0$, $x_{12} = 230$, $x_{13} = 70$, $x_{14} = 0$, $x_{21} = 0$, $x_{22} = 0$, $x_{23} = 0$, $x_{24} = 270$, $x_{31} = 150$, $x_{32} = 0$, $x_{33} = 10$, $x_{34} = 20$ and could be summarized as

	Consumer #1	Consumer #2	Consumer #3	Consumer #4	Supplier total
Supplier #1	0	230	70	0	300
Supplier #2	0	0	0	270	270
Supplier #3	150	0	10	20	170
Consumer total	150	230	80	290	**Total transportation cost: $13,550**

4.2   Nonlinear Programming: Gradient

Gradient of a function of several variables, $Q(x_1, x_2, \ldots, x_n)$, is defined as a vector comprising partial derivatives of this function with respect to individual variables, i.e.

$$\nabla Q(X) = \nabla Q(x_1, x_2, \ldots, x_n) = \begin{bmatrix} \frac{\partial Q}{\partial x_1} & \frac{\partial Q}{\partial x_2} & \cdots & \frac{\partial Q}{\partial x_n} \end{bmatrix}^{\mathrm{T}}$$

The above expression refers to an analytical definition of the gradient, however, it could be numerically defined at a particular location of the problem space, $X^* = [x1^*, x2^*, \ldots, xn^*]^T$. Let us refer to numerically defined gradient as $\nabla Q(X^*)$ where * is the index of the particular point where this gradient is defined. It is known that a numerically defined gradient is a good navigational tool: it is a vector always pointing in the direction of the increase of function Q in the space X.

Let us utilize this property of gradient for the minimization of function $Q(X)$. First, select some initial point $X^1 = [x_1^1, x_2^1, \ldots, x_n^1]^T$ and numerically evaluate derivatives of function $Q(X)$ in the vicinity of this point:

$$\frac{\partial Q(X^1)}{\partial x_1} \approx \frac{Q(x_1^1 + \Delta,\ x_2^1,\ \ldots, x_n^1) - Q(x_1^1,\ x_2^1,\ \ldots, x_n^1)}{\Delta}$$

$$\frac{\partial Q(X^1)}{\partial x_2} \approx \frac{Q(x_1^1,\ x_2^1 + \Delta,\ \ldots, x_n^1) - Q(x_1^1,\ x_2^1,\ \ldots, x_n^1)}{\Delta}$$

$$\cdots\cdots\cdots\cdots\cdots\cdots\cdots\cdots$$

$$\frac{\partial Q(X^1)}{\partial x_i} \approx \frac{Q(x_1^1,\ x_2^1,\ \ldots,\ x_i^1 + \Delta,\ \ldots, x_n^1) - Q(x_1^1,\ x_2^1,\ \ldots, x_n^1)}{\Delta}$$

$$\cdots\cdots\cdots\cdots\cdots\cdots\cdots\cdots$$

$$\frac{\partial Q(X^1)}{\partial x_n} \approx \frac{Q(x_1^1,\ x_2^1,\ \ldots,\ x_n^1 + \Delta) - Q(x_1^1,\ x_2^1,\ \ldots, x_n^1)}{\Delta}$$

where Δ is a small positive increment chosen on the basis of experience and intuition (V.S.: $\Delta = 0.0001$ is a good choice). Note that this approximation of derivatives, known as a forward difference, is not unique but is good enough for most applications. Now, when the direction towards the increase of function $Q(X)$ is known, and the direction towards the minimum is the opposite one, we can make a step from the initial point X^1 to the new point X^2 that is expected to be closer to the point of minimum: $X^2 = X^1 - a \cdot \nabla Q(X^1)$. Individual components of point X^2 will be defined as follows:

$$x_1^2 = x_1^1 - a \cdot \frac{\partial Q(X^1)}{\partial x_1}$$

$$x_2^2 = x_2^1 - a \cdot \frac{\partial Q(X^1)}{\partial x_2}$$

$$\cdots\cdots\cdots\cdots\cdots\cdots$$

$$x_n^2 = x_n^1 - a \cdot \frac{\partial Q(X^1)}{\partial x_n}$$

Now the procedure will repeat itself, but derivatives will be calculated in the vicinity of the new point X^2 and a transition to the point X^3 will be performed. This iterative process will lead to the vicinity of the minimum point of function $Q(X)$ providing that some conditions be met. Parameter a in the expressions above is a positive adjustable constant responsible for the convergence rate of the minimization procedure. Its initial value is arbitrarily defined and could be changed (typically decreased) in the process according to the following rule. Assume that the transition from point X^k to X^{k+1} is taking place: $X^{k+1} = X^k - a \cdot \nabla Q(X^k)$. The transition is successful if $Q(X^{k+1}) < Q(X^k)$, however in the situation when $Q(X^{k+1}) \geq Q(X^k)$

the value of parameter a must be reduced, for example by half, and the transition must be repeated with the value $a^{NEW} = 0.5a$, i.e. $X^{k+1} = X^k - a^{NEW} \cdot \nabla Q(X^k)$. If necessary, a value shall be repeatedly reduced until a successful transition will take place. The reduced a value shall be kept unchanged for the consequent step.

Termination conditions for the described procedure could be defined in a number of ways. First, and the simplest, is the definition of the maximum number of iterations (successful reduction steps of the function to be minimized). It is also common to stop the procedure if several (5, 10, 20) iterations did not result in a noticeable change in the optimization variables, i.e. $\left| X^{k-6} - X^{k-5} \right| \leq \xi$ and $\left| X^{k-5} - X^{k-4} \right| \leq \xi$ and ... and $\left| X^{k+1} - X^k \right| \leq \xi$ where $\xi > 0$ is a small arbitrary number. A block diagram of the procedure is seen in Fig. 4.8.

Fig. 4.8 Block diagram of gradient minimization procedure

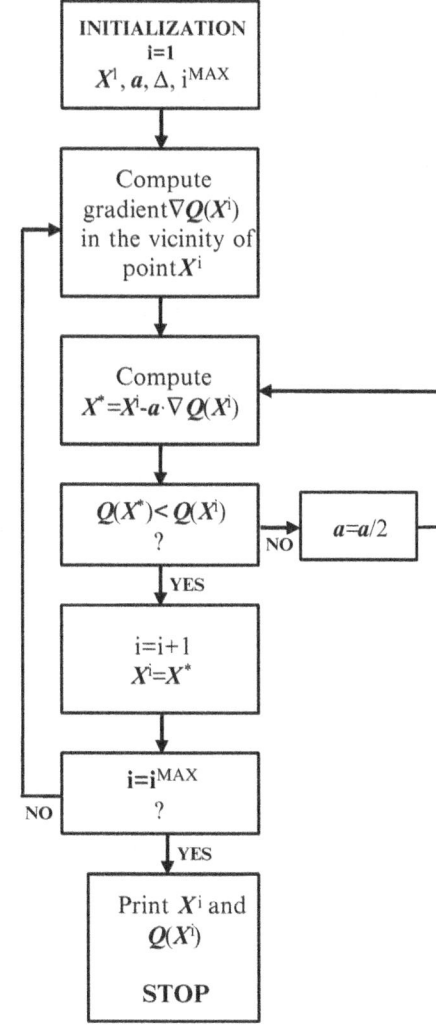

The gradient minimization procedure is quite common due to its simplicity. It does not require analytical expressions for derivatives. Values of function Q may be defined by analytical expressions or experimentally. The drawbacks of this approach are also evident. The function must be continuous, otherwise working with derivatives presents an impossible task. This reality creates difficulties with constrained minimization. The approach implies that the function to be minimized has only one minimum point: it works only as a local minimization technique.

4.3 Nonlinear Programming: Search

Search-based optimization presents a valuable alternative to gradient optimization: it does not utilize derivatives of the function to be optimized thus expanding the range of its applications to discontinuous functions. But, how common are the discontinuous functions? It is common to introduce constraints in the optimization procedure through so-called penalty functions, and penalty functions are the typical sources of discontinuities. Therefore, search becomes very useful in many practical problems.

4.3.1 Penalty Functions

Consider the following optimization problem where criterion and constraints are represented by generally speaking, nonlinear functions $Q(.)$ and $f_i(.)$, $i = 1, 2 \ldots$:

$$\text{Minimize } Q(x_1, x_2, \ldots x_n)$$

subject to conditions

$$f_1(x_1, x_2, \ldots x_n) \geq a_1$$
$$f_2(x_1, x_2, \ldots x_n) \geq a_2$$
$$\ldots\ldots\ldots\ldots\ldots\ldots\ldots\ldots\ldots$$
$$f_K(x_1, x_2, \ldots x_n) \geq a_K$$

Introduce penalty functions defined as

$$P_i(x_1, x_2, \ldots, x_n) = \begin{cases} C_i \cdot [f_i(x_1, x_2, \ldots, x_n) - a_i]^2, & \text{if } f_i(x_1, x_2, \ldots, x_n) \geq a_i \\ 0, & \text{if } f_i(x_1, x_2, \ldots, x_n) < a_i \end{cases}$$

$$\text{or } P_i(x_1, x_2, \ldots, x_n) = \begin{cases} C_i \cdot |f_i(x_1, x_2, \ldots, x_n) - a_i|, & \text{if } f_i(x_1, x_2, \ldots, x_n) \geq a_i \\ 0, & \text{if } f_i(x_1, x_2, \ldots, x_n) < a_i \end{cases}$$

$$\text{or } P_i(x_1, x_2, \ldots, x_n) = \begin{cases} C_i, & \text{if } f_i(x_1, x_2, \ldots, x_n) \geq a_i \\ 0, & \text{if } f_i(x_1, x_2, \ldots, x_n) < a_i \end{cases}$$

where $C_i \gg 1$ are arbitrary weights reflecting the importance of particular constraints, $i = 1, 2, \ldots K$. Then the original constrained optimization problem can be represented by the following unconstrained optimization problem

$$\text{Minimize } L(x_1, x_2, \ldots, x_n) = Q(x_1, x_2, \ldots, x_n) + \sum_{i=1}^{K} P_i(x_1, x_2, \ldots, x_n)$$

Function $L(.)$ is commonly referred to as the "loss function". It could be seen that due to the definition of penalty functions $P_i(.)$ it is a discontinuous function. It also could be seen that due to large values of weights C_i virtually any minimization algorithm would first "drive" penalty values to zero, and then, when constraints are satisfied, minimize the original function $Q(.)$.

Consider the following example illustrating the introduction of penalty functions.

Example 4.6 unconstrained optimization problem

Minimize $Q(x_1, x_2, x_3) = 5(x_1 + 6)^2 + 2(x_1 \cdot x_2 - 6x_3)^2 - 10x_2(x_3 - 2)^3$
subject to conditions:

$$x_1 + x_2 + 6x_3 = 10$$
$$0 \leq x_1 \leq 25$$
$$-10 \leq x_2 + x_3 \leq 10$$
$$x_1 - 4x_3 \leq 100$$

Define penalty functions representing the imposed constraints:

$$P_1 = 10^{15} \cdot [x_1 + x_2 + 6x_3 - 10]^2$$

$$P_2 = \begin{cases} 10^{10} \cdot x_1^2, & \text{if } x_1 < 0 \\ 0, & \text{if } x_1 \geq 0 \end{cases}$$

$$P_3 = \begin{cases} 10^{10} \cdot (x_1 - 25)^2, & \text{if } x_1 > 25 \\ 0, & \text{if } x_1 \leq 0 \end{cases}$$

$$P_4 = \begin{cases} 10^{10} \cdot (x_2 + x_3 - 10)^2, & \text{if } (x_2 + x_3 - 10)^2 > 0 \\ 0, & \text{otherwise} \end{cases}$$

$$P_5 = \begin{cases} 10^{10} \cdot (x_1 - 4x_3 - 100)^2, & \text{if } x_2 - 4x_3 > 100 \\ 0, & \text{otherwise} \end{cases}$$

The resultant loss function

$$L(x_1, x_2, x_3) = 5(x_1 + 6)^2 + 2(x_1 \cdot x_2 - 6x_3)^2 - 10x_2(x_3 - 2)^3 + \sum_{i=1}^{5} P_i(x_1, x_2, x_3)$$

could be easily defined by a computer code. Understandably, it should be minimized by a procedure that does not utilize derivatives

$$\frac{\partial L(x_1, x_2, x_3)}{\partial x_1}, \quad \frac{\partial L(x_1, x_2, x_3)}{\partial x_2}, \quad \frac{\partial L(x_1, x_2, x_3)}{\partial x_3}$$

It should also be noted that due to nonlinear criterion and constraints, this problem most likely does not have one minimum, and finding the global minimum presents an additional challenge. As it is commonly done when some parameters are arbitrarily chosen (in this case, weight coefficients) the user shall inspect the obtained solution and if necessary, change the weight values. It is a good practice to demonstrate that the solution does not depend on the choice of the weights.

4.3.2 Random Search

This approach could be perceived as the most straight forward "trial-and-error" technique utilizing the full power of a modern computer and perhaps a supercomputer. It facilitates finding the global solution of linear and nonlinear, constrained and unconstrained, continuous and discontinuous optimization problems. Its only drawback is the gigantic amount of computations that is prohibitive in many practical situations. The strategy of random search is illustrated by Fig. 4.9.

4.3.3 Simplex Method of Nelder and Mead

Direct search is a much more efficient alternative to random search. One can define direct search as a thoughtful and insightful trial-and-error approach. It still has to start from some initial conditions but its steps are based on a reasonable expectation of success. It works well with continuous and discontinuous, linear and nonlinear, constrained and unconstrained functions. Its only drawback compared to random search is the inherent inability to assure that the global minimum be found. This fault is not that crucial: direct search is typically used in realistic situations where properly chosen, the initial point guarantees that the global minimum can be found. Since direct search does not call for a gigantic number of steps, it could be used in situations when values of the objective functions are defined by computer simulations and even by physical experiments.

Although there is a good number of direct search procedures utilizing different rationale for making the "next step," one of the most practical is the Simplex Method by Nelder-Mead (1965). The algorithm works with $n + 1$ vertices of a simplex (convex polytope) defined in the n-dimensional search space. It calculates (obtains) numerical values of the function to be minimized at every vertex, compares these values, and implements some rules for replacing the worst vertex (i.e.

Fig. 4.9 Random search

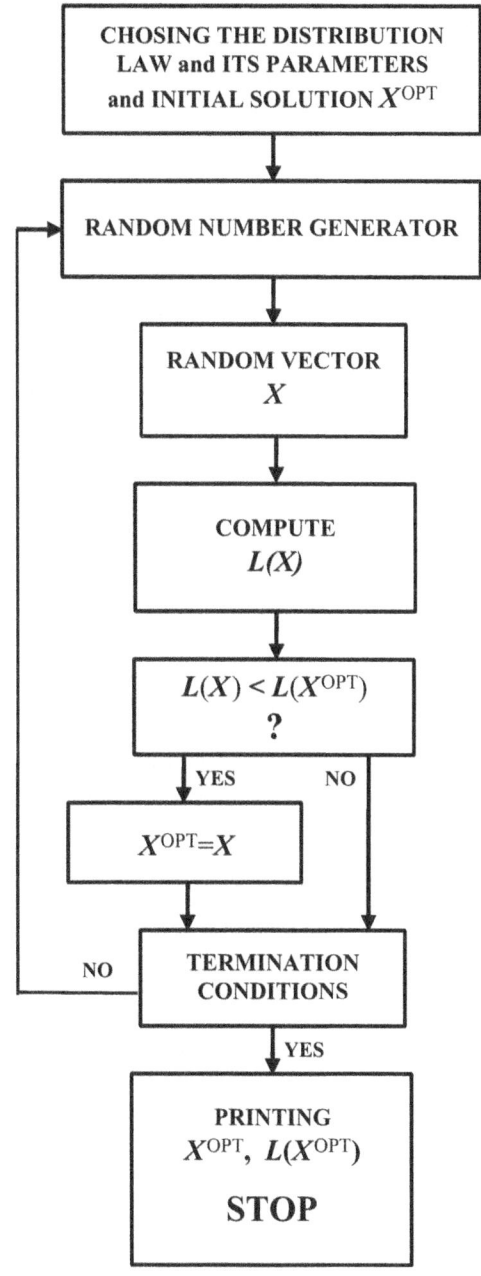

the one with the largest value of the objective function). This process could be best illustrated in two dimensional space when simplex, with its three vertices, is just a triangle.

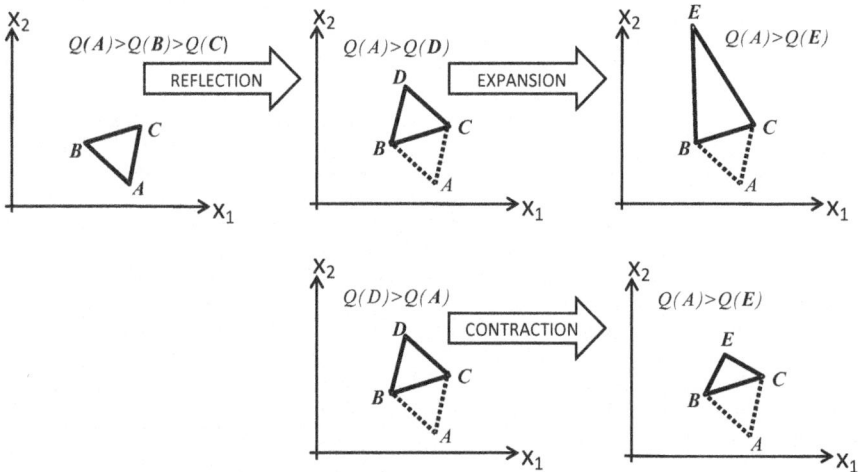

Fig. 4.10 How the simplex procedure works

First assume that the initial simplex with vertices A, B, C is established. It is often done by specifying some initial point, say point A, and the step size that determines the size of the resultant initial simplex, i.e. triangle ABC. Next is the evaluation of the objective function $Q(x_1,x_2)$ at each vertex (x_1, x_2 are coordinates of points A, B, C) thus resulting in numerical values $Q(A)$, $Q(B)$ and $Q(C)$. Assume that the comparison reveals that $Q(A) > Q(B) > Q(C)$, and since our task is minimization, the "worst" point is A. Then as seen in Fig. 4.10 above, the algorithm performs a special operation, reflection, thus establishing a new point D. What happens next, depends on the value $Q(D)$. If $Q(A) > Q(D)$, the algorithm performs expansion as shown above, creating a new point E. The expansion could be repeated providing that still $Q(A) > Q(E)$. In the situation when $Q(D) > Q(A)$, the contraction is performed. It should be performed repeatedly until condition $Q(A) > Q(E)$ is achieved. Upon the establishment of the "new" point E, the "old" point A is discarded. Now the new simplex with vertices B, C, and E is ready for performing the same computational cycle.

The termination conditions can be defined in terms of the total number of steps (optimization cycles), or in terms of the distance between vertices of the simplex.

It is good to realize that besides "purely computational" applications, the Simplex procedure can be implemented in the "(wo)man in the loop" regime for the real-time optimization of technical systems that could be represented by a simulator. Figure 4.11 below illustrates an application of the Simplex optimization to the tuning of a PID (proportional-integral-derivative) controller. The Vissim-based simulator (see http://www.vissim.com/) features a controlled process with a PID controller with manually adjustable parameters K_P, K_I, and K_D known as proportional, integral and derivative gains.

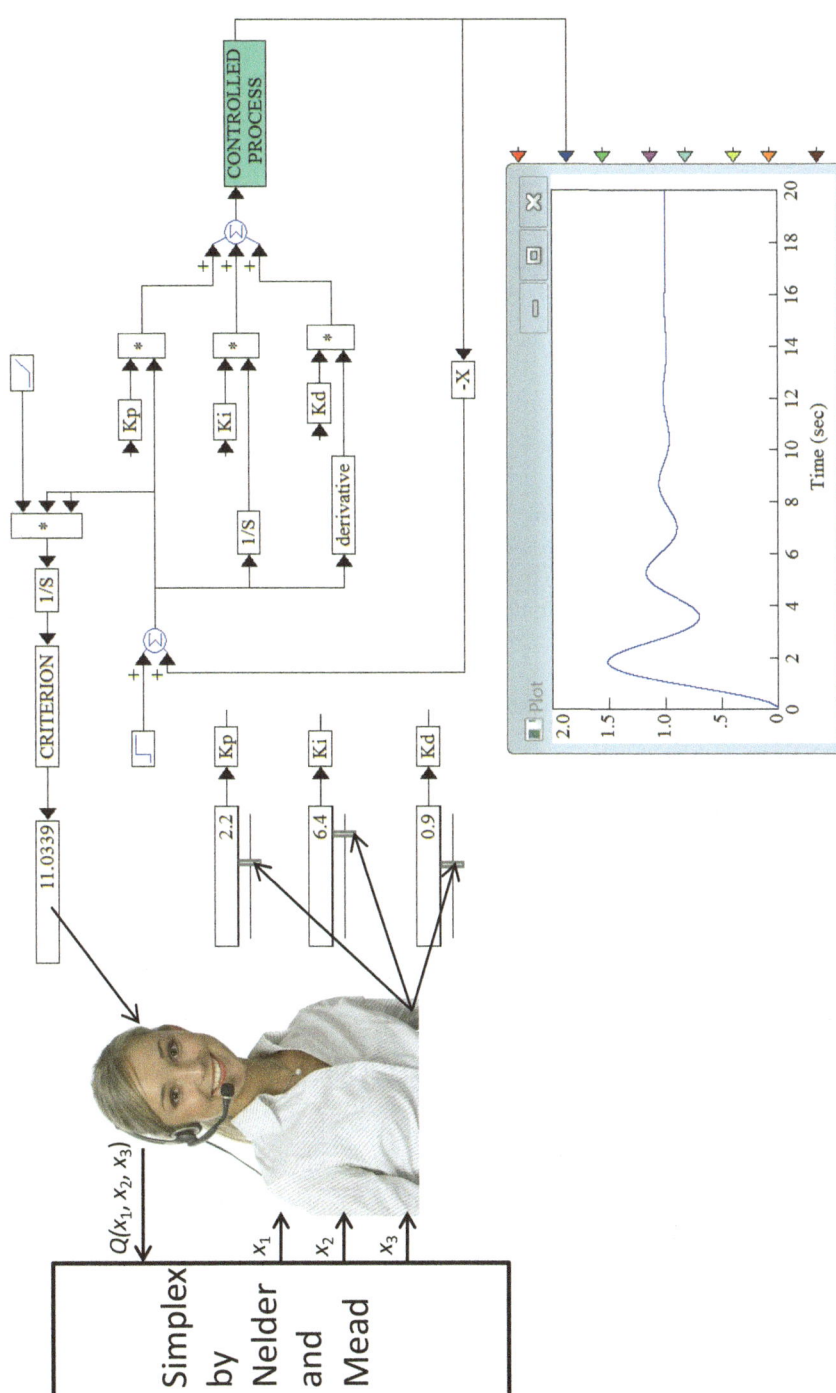

Fig. 4.11 Simplex application to tuning PID Controller

The optimization criterion is the commonly used ITSE (integral-time-squared-error) defined as

$$Q(K_P, K_I, K_D) = \int_0^T t \cdot e^2 \cdot dt$$

where e is the system error (the discrepancy between the actual and desired system output values), t is continuous time, and T is the simulation period. It is known from Controls that minimization of ITSE-type criteria leads to the most desirable transient process in the system.

4.3.4 Exercise 4.1

Problem 1 Solving a mixing problem. The table below contains characteristics of several materials that are to be mixed to obtain a raw material for a metallurgical process. Obtain the mixture recipe that would have the following required chemical composition and total volume at minimum cost. The mixture characteristics are as follows:

Fe $\geq 20\,\%$, Zn $\geq 10\,\%$, SiO$_2 \leq 42\,\%$, Cu $\geq 5\,\%$, total weight 500 tons

	Fe %	Zn %	SiO$_2$ %	Cu %	Cost, \$/ton	Availability
Material 1	15	38	41	6	120	250 tons
Material 2	40	12	40	1	150	590 tons
Material 3	35	5	27	28	211	1000 tons
Material 4	16	11	21	18	140	520 tons
Material 5	33	1	60	5	75	2500 tons
Material 6	7	23	45	25	214	800 tons

Problem 2 Solving an LSM parameter estimation problem using a gradient procedure. Generate input and the output variables as follows (k = 1, 2,..., 500):

$$x_1(k) = 5 + 3 \cdot \text{Sin}\,(17 \cdot k) + \text{Sin}\,(177 \cdot k) + .3 \cdot \text{Sin}\,(1771 \cdot k)$$

$$x_2(k) = 1 - 2 \cdot \text{Sin}\,(91 \cdot k) + \text{Sin}\,(191 \cdot k) + .2 \cdot \text{Sin}\,(999 \cdot k)$$

$$x_3(k) = 3 + \text{Sin}\,(27 \cdot k) + .5 \cdot \text{Sin}\,(477 \cdot k) + .1 \cdot \text{Sin}\,(6771 \cdot k)$$

$$x_4(k) = -.1 \cdot x_1(k) + .3 \cdot x_2(k) + 2.5 \cdot \text{Sin}\,(9871 \cdot k) + .7 \cdot \text{Cos}(6711 \cdot k)$$

$$y(k) = 2 \cdot x_1(k) + 3 \cdot x_2(k) - 2 \cdot x_3(k) + 5 \cdot x_4(k) + .3 \cdot \text{Sin}\,(1577 \cdot k) + .2 \cdot \text{Cos}(7671 \cdot k)$$

Obtain "unknown" coefficients of the regression equation

$$y^{MOD}(k) = a_1 x_1(k) + a_2 x_2(k) + a_3 x_3(k) + a_4 x_4(k)$$

using the least squares method implemented via the gradient procedure listed below (that could be rewritten in MATLAB). Assume zero initial values of the coefficients. Compute the coefficient of determination of the obtained regression equation.

Problem 3 Utilize data of Problem #2 to obtain coefficients of the regression equation $y^{MOD}(k) = a_1 x_1(k) + a_2 x_2(k) + a_3 x_3(k) + a_4 x_4(k)$ applying the gradient procedure. It is required, however, that all regression coefficients be positive. Show the obtained coefficients. Compute the coefficient of determination for the resultant regression equation. Explain the change in the coefficient of determination comparing with Problem #2

```
PROGRAM GRADIENT
  DIMENSION X(10),X1(10),DER(10)
  WRITE(*,*)' ENTER NUMBER OF VARIABLES '
  READ(*,*) N
  WRITE(*,*)' ENTER THE gain OF THE PROCEDURE '
  READ(*,*)A
  WRITE(*,*)' ENTER INITIAL NUMBER OF STEPS '
  READ(*,*) NSTEP
  H=.001
  DO 1 I=1,N
  WRITE(*,*)' ENTER INITIAL VALUE FOR X(',I,')'
1   READ(*,*)X(I)
10  CONTINUE
  K=1
  CALL SYS(N,X,Q)
  QI=Q
100 CONTINUE
  DO 4 I=1,N
  X(I)=X(I)+H
  CALL SYS(N,X,Q1)
  DER(I)=(Q1-Q)/H
  X(I)=X(I)-H
4   CONTINUE
50  CONTINUE
  DO 5 I=1,N
5   X1(I)=X(I)-DER(I)*A
  CALL SYS(N,X1,Q1)
  IF(Q1.GE.Q) A=A/2
  IF(Q1.GE.Q) GOTO 50
  DO 30 I=1,N
30  X(I)=X1(I)
  Q=Q1
  IF(ABS(Q).LE.1e-5)GOTO 2
  K=K+1
  IF(K.GT.NSTEP) GOTO 2
```

```
      GOTO 100
2     CONTINUE
      WRITE(*,*)' ITERATIONS RUN: ',NSTEP
      WRITE(*,*)' INITIAL CRITERION AVLUE: ',QI
      WRITE(*,*)' CRITERION VALUE REACHED: ',Q
      DO 7 I=1,N
7     WRITE(*,*)' OPTIMAL VALUE: X(',I,') = ',X(I)
      WRITE(*,*)' ENTER ADDITIONAL NUMBER OF STEPS '
      IF(ABS(Q).LE.1e-5)CALL EXIT
      READ(*,*) NSTEP
      IF(NSTEP.EQ.0)CALL EXIT
      GOTO 10
      END
C
      SUBROUTINE SYS(N,X,Q)
      DIMENSION X(10)
      Q=0.
      DO 1 I=1,N
      Q=Q+(X(I)-5.*I)**2
1     CONTINUE
      Q=Q**2
      RETURN
           END
```

4.4 Genetic Optimization

Genetic optimization algorithms possess the advantages of random and direct search optimization procedures. Combined with the availability of high performance computers they alleviate major obstacles in the way of solving multivariable, nonlinear constrained optimization problems. It is believed that these algorithms emulate some concepts of the natural selection process responsible for the apparent perfection of the natural world. One can argue about the concepts, but the terminology of genetic optimization is surely adopted from biological sciences.

Assume that we are in the process of finding the optimum, say the maximum, of a complex, multivariate, discontinuous, nonlinear cost function $Q(X)$. The constraints of the problem have already been addressed by the penalty functions introduced in the cost function and contributing to its complexity.

Introduce the concepts of an individual, generation, and successful generation. An *individual* is an entity that is characterized by its location in the solution space, X^I and the corresponding value of the function Q, i.e. $Q(X^I)$. A *generation* is a very large number of individuals created during the same cycle of the optimization procedure. A *successful generation* is a relatively small group of K individuals that have some common superior trait, for example, they all have the highest associated values $Q(.)$ within their generation. The genetic algorithm consists of repeated cycles of creation of successful generations.

Creation of the Initial Generation First, the feasibility range $[x_k^{MIN}, x_k^{MAX}]$ for each solution variable x_k $k = 1,2,3,\ldots,n$, is to be established. Each interval $[x_k^{MIN}, x_k^{MAX}]$ is divided into the same number of subintervals, say L, thus resulting in a grid within the solution space with numerous nodes. The next task is the evaluation of function Q at each node of the grid, i.e. the creation of individuals "residing" at every node. During this process the successful generation is selected consisting of K individuals that have the highest values of the function Q. It is done by forming a group of individuals ordered according to their Q values, i.e.

$$Q(X^K) \leq Q(X^{K-1}) \leq \ldots Q(X^2) \leq Q(X^1) \quad (*)$$

Any newly generated individual X^I is discarded if $Q(X^I) \leq Q(X^K)$. However if $Q(X^I) > Q(X^L)$, it is included in the group replacing the individual X^L with the lowest Q value. Therefore the successful generation still includes K individuals that are being renumbered and reordered to assure (*). This process is repeated each time a new individual is generated i.e. until the entire initial generation is created and analyzed.

Creation of the Next Successful Generation involves only members of the existing successful generation. Two techniques are utilized for this purpose, parenting and mutation. Parenting (crossover) involves two individuals, X^A and X^B and results in an "offspring"

$$X^C = \left[x_1^C, x_2^C, \ldots x_k^C, \ldots x_n^C \right]^T$$

defined as follows:

$$x_1^C = \lambda_1 x_1^A + (1 - \lambda_1) x_1^B$$
$$x_2^C = \lambda_2 x_2^A + (1 - \lambda_2) x_2^B$$
$$\ldots\ldots\ldots$$
$$x_k^C = \lambda_k x_k^A + (1 - \lambda_k) x_k^B$$
$$\ldots\ldots\ldots$$
$$x_n^C = \lambda_n x_n^A + (1 - \lambda_n) x_n^B$$

where $0 < \lambda_k < 1$ are random numbers generated by a random number generator. Then, based on the computation of $Q(X^C)$ the newly created individual X^C is accepted into the successful generation or discarded. The parenting process is repeated several number times for every combination of two members of the original successful generation.

The mutation process implies that every member of the original successful generation, X^I originates a "mutant" $X^M = [x_1^M, x_2^M, \ldots x_k^M, \ldots x_n^M]^T$ defined as follows:

$$x_1{}^M = \alpha_1 x_1{}^I$$
$$x_2{}^M = \alpha_2 x_2{}^I$$
$$\dots\dots\dots\dots$$
$$x_k{}^M = \alpha_k x_k{}^A$$
$$\dots\dots\dots\dots$$
$$x_n{}^M = \alpha_n x_n{}^A$$

where α_k are normally distributed random numbers generated by a random number generator. Based on the computation of $Q(X^M)$ the newly created individual X^M is accepted into the successful generation or discarded. The mutation process is repeated several number times for every member of the original successful generation.

Understandably, parenting and mutation upon completion results in a new successful generation that is to be subjected to a new cycle of the procedure unless the termination conditions be satisfied. The most common termination condition refers to the variability within a successful generation, and could be expressed as:

$$\sum_{i=1}^{K-1} \left| X^i - X^{i+1} \right| \leq \delta$$

where $\delta > 0$ is some judiciously chosen small positive number.

It is good to remember that genetic optimization is capable of finding a global minimum of virtually any function $Q(X)$. Moreover, it works even when this function does not exist as an analytical expression: in this situation for any particular X^I the value of $Q(X^I)$ could be determined by running a computer simulation or by an experiment. Figure 4.12 provides a block diagram of the genetic optimization procedure.

The following MATLAB code implementing a genetic optimization procedure was written by my former student Dr. Jozef Sofka

```
%genetic algorithm for minimization of a nonlinear function
%(c) Jozef Sofka 2004
%number of crossovers in one generation
cross = 50;
%number of mutations in one generation
mut = 30;
%extent of mutation
mutarg1 = .5;
%size of population
population = 20;
%number of alleles
al = 5;
%trying to minimize function
%abs(a^2/b+c*sin(d)+b^c+1/(e+a)^2)
 clear pop pnew;
 %definition of "best guess" population
 pop(1:population,1) = 12+1*randn(population,1);
 pop(1:population,2) = 1.5+.1*randn(population,1);
 pop(1:population,3) = 13+1*randn(population,1);
```

Fig. 4.12 Block diagram of
a genetic optimization
procedure

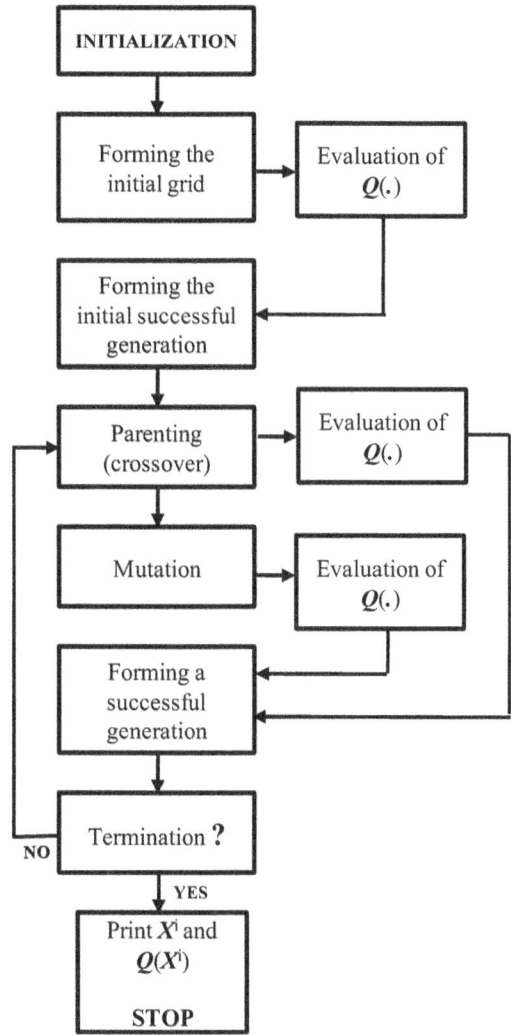

```
 pop(1:population,4) =1.5 + .2*randn(population,1);
 pop(1:population,5) = (.5*randn(population,1));
% evaluation of fitness population
for f =1:population
   e(f) =abs(pop(f,1)^2/pop(f,2) +pop(f,3)*sin(pop(f,4)) +pop(f,2)
^pop(f,3) +1/(pop(f,5) +pop(1))^2);
 end
[q,k] =sort(e);
%number of generations
for r=1:500
  parameters(r,1:al) =pop(k(1),1:al);
  fitness(r) =e(k(1));
%crossover
for f =1:cross
```

```
p1 = round((rand + rand)/2*(population-1)) + 1;
p2 = round((rand + rand)/2*(population-1)) + 1;
p3 = (2*rand-.5);
     pnew(f,:) = pop(k(p1),1:a1) + p3*(pop(k(p2),1:a1)-pop(k(p1),1:
a1));
 %evaluation of fitness
 fit(f) = abs(pnew(f,1)^2/pnew(f,2) + pnew(f,3)*sin(pnew(f,4)) + pnew
(f,2)^pnew(f,3) + 1/(pnew(f,5) + pnew(1))^2);
 end
%selection
for f = 1:cross
 if (fit(f) < e(k(population-3)))
  pop(k(population),:) = pnew(f,:);
  e(k(population)) = fit(f);
  [q,k] = sort(e);
  end
  end
%mutation
for f = 1:mut
 p = round(rand*(population-1)) + 1;
 o = round((a1-1)*rand) + 1;
 pnew(f,:) = pop(p,:);
 pnew(f,o) = pnew(f,o) + mutarg1*randn(1,1);
 %evaluation of fitness
 fit(f) = abs(pnew(f,1)^2/pnew(f,2) + pnew(f,3)*sin(pnew(f,4)) + pnew
(f,2)^pnew(f,3) + 1/(pnew(f,5) + pnew(1))^2);
 end
%selection
for f = 1:mut
 if (fit(f) < e(k(population-1)))
  pop(k(population),:) = pnew(f,:);
  e(k(population)) = fit(f);
  [q,k] = sort(e);      end
 end
end
fprintf('Parameters a = %f; b = %f; c = %f; d = %f; e = %f\n', ...,
pop(k(1),1), pop(k(1),2), pop(k(1),3), pop(k(1),4), pop(k(1),5))
fprintf('minimize   function   abs(a^2/b + c*sin(d) + b^c + 1/(e + a)^2)
\n')
figure
plot(parameters)
figure
  semilogy(fitness)
```

4.4.1 Exercise 4.2

Problem 1 Use Simplex Optimization procedure (to be provided) to tune parameters of a PID controller as shown in Fig. 3.3. The simulation setup could be implemented in Simulink or Vissim. The following transfer function is recommended for the controlled plant:

$$G(s) = \frac{s + 6}{s^3 + 6s^2 + 10s + 10}$$

To show the effectiveness of the tuning procedure provide a sequence (five or so) of numerical values of the parameters of the controller, values of the criterion, and the system step responses.

Problem 2 Given input–output data representing a highly nonlinear, static process:

x_1	x_2	x_3	y
1	1	1	17.59141
1	1	2	21.59141
1	2	2	44.94528
2	2	2	81.89056
2	2	3	89.89056
2	3	3	216.8554
3	3	3	317.2831
-3	3	3	-285.2831
-3	-3	3	15.25319
-3	-3	-3	-0.496806
-3	3	-3	-301.0331
-1	3	-3	-100.1777
-1	3	5	-36.42768
-5	2	4	-152.7264
5	2	1	188.7264

Given the configuration of the mathematical model of this process:

$$y^{MOD} = a_1 x_1 e^{a_2 x_2} + a_3{}^{(a_4 x_3 + a_5)}$$

Utilize the Genetic Optimization (GO) program provided above and the input/output data to estimate unknown parameters of the mathematical model given above. Experiment with values of the control parameters of the GO procedure. Compute the coefficient of determination for the obtained regression model and comment on the model accuracy. Document your work.

4.5 Dynamic Programming

Many physical, managerial, and controlled processes could be considered as a sequence of relatively independent but interrelated stages. This division, natural or imaginative, could be performed in the spatial, functional, or temporal domains. The following diagram in Fig. 4.13 represents a typical multi-stage process containing four stages. Every stage or sub-process is relatively independent in the sense that it is characterized by its own (local) input x_i, local output y_i, local control

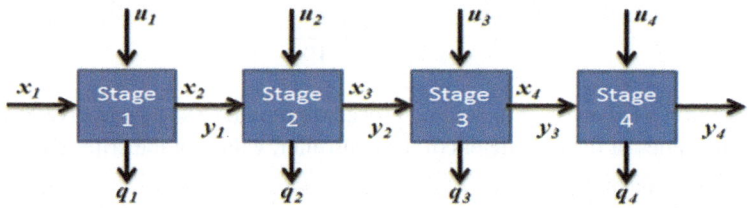

Fig. 4.13 Multi-stage process with four stages

effort u_i, and the local goodness criterion q_i. Both the output and the criterion of each stage (sub-process) are defined by its local input and the control effort, i.e. $y_i = y_i(x_i, u_i)$ and $q_i = y_i(x_i, u_i)$.

At the same time, individual stages (sub-processes) are interrelated. Indeed the output of every stage, except the last (n-th) stage, serves as the input of the consequent stage, i.e. for $i = 1, 2, 3, \ldots, n - 1$ $y_i = x_{i+1}$. This reality results in the following relationships that links the entire sequence:

$$y_i = y_i(x_i, u_i) = y_i[y_{i-1}(x_{i-1}, u_{i-1}), u_i] = y_i(x_{i-1}, u_{i-1}, u_i)$$
$$= y_i[y_{i-2}(x_{i-2}, u_{i-2}), u_{i-1}, u_i] = y_i(x_{i-2}, u_{i-2}, u_{i-1}, u_i) = \ldots$$
$$= y_i(x_1, u_1, u_2, u_3, \ldots, u_i)$$

and similarly $q_i = q_i(x_1, u_1, u_2, u_3, \ldots, u_i)$, where $i = 1,2,3,\ldots, n$ is the sequential number of the stage.

These relationships indicate that the output and criterion value of any stage of the process, except the first stage, are defined by the input of the first stage, control effort applied at this stage and control efforts applied at all previous stages. In addition to the above relationships, the stages of the process are linked by the "overall goodness criterion" defined as the sum of all "local" criteria,

$$Q = \sum_{k=1}^{n} q(x_k, u_i)$$ where n is the total number of the stages. It could be seen that

the overall criterion depends on the input of the first stage and all control efforts, i.e.

$$Q = Q(x_1, u_1, u_2, u_3, \ldots, u_n)$$

Therefore the optimization problem of a multistage process implies the minimization (maximization) of the overall criterion $Q(.)$ with respect to control efforts applied at individual stages, u_k, $k = 1, 2, \ldots n$, for any given input of the first stage, x_1, and may be subject to some constraints imposed on the outputs of the individual stages, y_k, $k = 1, 2, \ldots n$. One can realize that the process optimization problem cannot be solved by the independent optimization of the individual stages with respect to their "local" criteria, q_k, $k = 1, 2, \ldots n$. The optimal control strategy must be "wise": "local" optimization of any sub-process may result in such an output that will completely jeopardize the operation of the consequent stages thus causing poor operation of the entire multistage process. Therefore, optimization of any stage of a multi-stage process must take into account the consequences of this optimization for all consequent stages. Selection of any "local" control effort cannot be performed without assessing its impact on the overall criterion.

Dynamic programming is an optimization technique intended for the optimization of multi-stage processes. It is based on the fundamental principle of optimality of dynamic programming formulated by Richard Bellman. *A problem is said to satisfy the Principle of Optimality if the sub-solutions of an optimal solution of the problem are themselves optimal solutions for their sub-problems.* Fortunately, optimization problems of multi-stage processes do satisfy the Principle of Optimality that offers a powerful solution approach in the most realistic situations. The key to the application of the Principle of Optimality is in the following statement that is stemming from this principle: *any last portion of an optimal sequence of steps is optimal.*

Let us illustrate this principle using the chart below in Fig. 4.14 that presents a process comprising of 12 sequential stages divided into two sections, AB and BC. It is assumed that each j-th stage of this process is characterized by its "local" criterion, q_j. Assume that the overall criterion of the process is defined as the sum of local criteria: $Q = \sum_{j=1}^{12} q_j$

Let us define the sectional criteria for each of the two sections: $Q_{AB} = \sum_{j=1}^{5} q_j$ and $Q_{BC} = \sum_{j=6}^{12} q_j$. Assume that for every stage of the process some control effort is chosen, such that the entire combination of these control efforts, u_j^{OPT}, $j = 1, 2, \ldots, 12$, optimizes (minimizes) the overall process criterion Q. Then according to the principle of dynamic programming control efforts u_j^{OPT}, $j = 6, 7,$ $\ldots, 12$ optimize the last section of the sequence, namely BC, thus bringing criterion $Q_{BC} = \sum_{j=6}^{12} q_j$ to its optimal (minimal) value. At the same time, control efforts u_j^{OPT}, $j = 1, 2, \ldots, 5$ are not expected to optimize section AB of the process, thus criterion $Q_{AB} = \sum_{j=6}^{12} q_j$ could be minimized by a completely different combination of control efforts, say u_j^{ALT}, $j = 1, 2, \ldots, 5$.

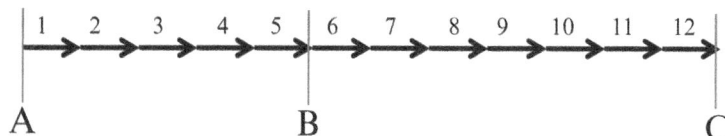

Fig. 4.14 Twelve stage process

The fundamental principle provides the framework for a highly efficient and versatile optimization procedure of dynamic programming that works on a step-by-step basis and defines optimal control efforts for individual stages of the multi-stage process. It is important that control decisions made at each step of the procedure do not optimize individual stages of the process, i.e. do not solve the "local" optimization problems. Instead, they optimize the *last portion* of the entire process that starts at the stage in question and end at the last stage of the process.

When doing so, every step of the optimization procedure takes into account not only the particular stage of the process but also all consequent stages. The procedure is iterative, therefore it shall start from the last section of the multistage process where there are no consequent stages to be considered. At the same time, the optimal solution of the control problem u_j^{OPT}, cannot be explicitly defined without knowing the input x_j applied to the appropriate section of the process. Therefore, the dynamic programming procedure is performed in two steps: conditional optimization and unconditional optimization. Conditional optimization starts from the end of the process addressing the last stage of the process first, then the last two stages of the process, then the last three stages, and finally the entire process. Why is it called conditional?—because at the first step, the procedure defines the optimal conditional control effort (OCCE) for the last stage of the process that is dependent on the input of the last stage of the process:

$$u_N^{OPT} = F(x_N)$$

that minimizes the sectional criterion

$$Q_N(x_N, u_N) = q_N(x_N, u_N)$$

(Note that the sectional criterion is marked by the index of the first stage of the section). Now the output of the last stage of the process (and the output of the entire process) is

$$y_N = y_N(x_N, u_N^{OPT})$$

This solution must be consistent with the required (allowed) value of the output of the process, Y^*, i.e.

$$y_N = y_N(x_N, u_N^{OPT}) = Y^*$$

At the next step the OCCE for the second stage from the end of the process is defined as a function of the input applied to this stage:

$$u_{N-1}^{OPT} = F(x_{N-1})$$

that minimizes the sectional criterion for the last two stages of the process:

$$Q_{N-1}\left(x_{N-1}, u_{N-1}, u_N^{OPT}\right) = q_{N-1}(x_{N-1}, u_{N-1}) + q_N\left(x_N, u_N^{OPT}\right)$$

Note that in this expression x_N does not work as an independent factor, it is defined as the output of the previous stage of the process:

$$x_N = y_{N-1}\left(x_{N-1}, u_{N-1}^{OPT}\right)$$

and therefore criterion Q_{N-1} actually depends only on two variable factors, x_{N-1} and u_{N-1}:

$$Q_{N-1}(x_{N-1}, u_{N-1}) = q_{N-1}(x_{N-1}, u_{N-1}) + q_N\left[y_{N-1}(x_{N-1}, u_{N-1}), u_N^{OPT}\right]$$

The solution must ensure that the resultant output

$$y_{N-1} = y_{N-1}\left(x_{N-1},\ u_{N-1}^{OPT}, u_N^{OPT}\right)$$

is within its allowed limits, i.e.

$$y_{N-1}^{\ \ MIN} \le y_{N-1} \le y_{N-1}^{\ \ MAX}$$

Now let us define the OCCE for the third stage from the end of the process as a function of the input applied to this stage:

$$u_{N-2}^{OPT} = F(x_{N-2})$$

that minimizes the sectional criterion that "covers" the last three stages:

$$Q_{N-2}\left(x_{N-2}, u_{N-2}, u_{N-1}^{OPT}, u_N^{OPT}\right) = q_{N-2}(x_{N-2}, u_{N-2}) + q_{N-1}\left(x_{N-1}, u_{N-1}^{OPT}\right)$$
$$+ q_N\left(x_N, u_N^{OPT}\right)$$

Again, in this expression x_{N-1} and x_N are not independent factors, they are defined as outputs of the previous stages of the process:

$$x_{N-1} = y_{N-2}(x_{N-2}, u_{N-2}) \text{ and } x_N = y_{N-1}\left(x_{N-1}, u_{N-1}^{OPT}\right)$$

and therefore criterion Q_{N-2} actually depends only on two variable factors, x_{N-2} and u_{N-2}

$$Q_{N-2} = q_{N-2}(x_{N-2}, u_{N-2}) + q_{N-1}\left[y_{N-2}\left(x_{N-2}, u_{N-2}^{OPT}\right), u_{N-1}^{OPT}\right]$$
$$+ q_N\left[y_{N-1}\left(x_{N-1}, u_{N-1}^{OPT}\right), u_N^{OPT}\right] = Q_{N-2}\left(x_{N-1}, u_{N-2},\ u_{N-1}^{OPT}, u_N^{OPT}\right)$$

The optimal value of this criterion is:

$$Q_{N-2}\left(x_{N-1}, u_{N-2}^{OPT}, u_{N-1}^{OPT}, u_N^{OPT}\right)$$

Again, the appropriate output,

$$y_{N-2} = y_{N-2}\left(x_{N-2}, u_{N-2}^{OPT}\right)$$

must be consistent with the allowed value for the output of the appropriate stage of the process:

$$y_{N-2}{}^{MIN} \leq y_{N-2} \leq y_{N-2}{}^{MAX}$$

It could be seen that eventually the procedure defines the control effort for the first stage of the process as a function of the input applied to this stage:

$$u_1^{OPT} = F(x_1)$$

that minimizes the sectional criterion

$$Q_1\left(x_1, u_1, u_2^{OPT}, u_3^{OPT}, \ldots, u_{N-1}^{OPT}, u_N^{OPT}\right)$$

However, the input of the first stage (the input of the overall process), x_1, is explicitly known, therefore the control effort u_1^{OPT} could be explicitly defined. This results in the explicitly defined output of the first stage, y_1. Since $x_2 = y_1$, the optimal conditional control effort

$$u_2^{OPT} = F(x_2)$$

could be explicitly defined thus resulting in an explicit definition of the output of the second stage and the input of the third stage, and so on... It could be seen that the procedure moves now from the first stage of the process to the last stage, converting conditional control efforts into explicitly defined unconditional optimal control efforts.

Let us consider the application of the outlined approach to the following numerical example representing the so-called optimal routing problem.

Example 4.7 Apply dynamic programming to establish the optimal ("minimum cost") route within the following graph in Fig. 4.15

It could be seen that the transportation problem featured by the above graph consists of five stages and four steps. Step 1 consists of four alternative transitions: $1/1 \rightarrow 2/1$, $1/1 \rightarrow 2/2$, $1/1 \rightarrow 2/3$ and $1/1 \rightarrow 2/4$ with the associated costs of 5, 3, 1, and 2 (units). Step 2 consists of 12 alternative transitions: $2/1 \rightarrow 3/1$, $2/1 \rightarrow 3/2$, $2/$

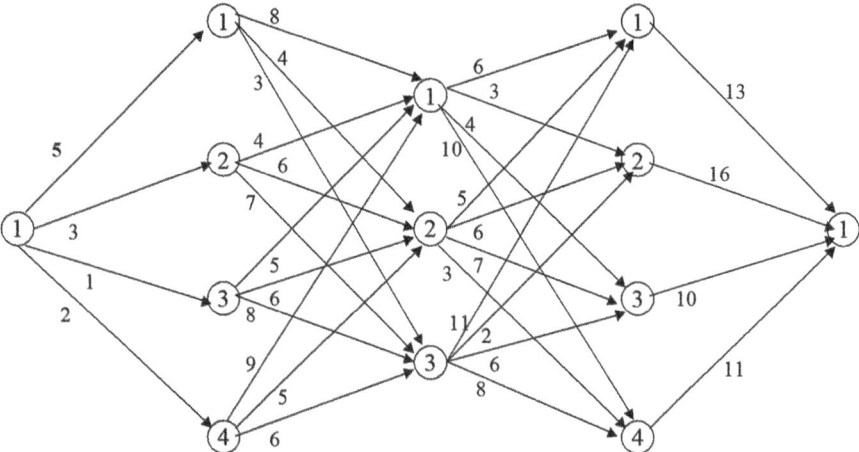

Fig. 4.15 Process graph

$1 \rightarrow 3/3$ with the associated costs of 8, 4, and 3 (units); $2/2 \rightarrow 3/1$, $2/2 \rightarrow 3/2$, $2/2 \rightarrow 3/3$ with the associated costs of 4, 6, and 7 (units); $2/3 \rightarrow 3/1$, $2/3 \rightarrow 3/2$, $2/3 \rightarrow 3/3$ with the associated costs of 5, 6, and 8 (units); and $2/4 \rightarrow 3/1$, $2/4 \rightarrow 3/2$, $2/4 \rightarrow 3/3$ with the associated costs of 9, 5, and 6 (units). Step 3 also consists of 12 alternative transitions: $3/1 \rightarrow 4/1$, $3/1 \rightarrow 4/2$, $3/1 \rightarrow 4/3$, $3/1 \rightarrow 4/4$ with the associated costs of 6, 3, 4, and 10 (units); $3/2 \rightarrow 4/1$, $3/2 \rightarrow 4/2$, $3/2 \rightarrow 4/3$, $3/2 \rightarrow 4/4$ with the associated costs of 5, 6, 7, and 3 (units); $3/3 \rightarrow 4/1$, $3/3 \rightarrow 4/2$, $3/3 \rightarrow 4/3$, $3/3 \rightarrow 4/4$ with the associated costs of 11, 2, 6, and 8 (units). Finally, the last step, 4, consists of four alternative transitions: $4/1 \rightarrow 5/1$, $4/2 \rightarrow 5/1$, $4/3 \rightarrow 5/1$ and $4/4 \rightarrow 5/1$ with the associated costs of 13, 16, 10, and 11 (units). It is required to establish such a sequence of transitions (optimal path) that would lead from the initial to the final stage (nodes of the above graph) and had the minimal sum of the transition costs.

Could we have established the optimal path by considering all possible alternative paths within this graph?—perhaps, but the required computational effort is expected to be very high. Should the number of stages and alternative transitions at every step be greater, this approach will become prohibitively formidable.

According to the dynamic programming procedure, let us define conditionally optimal transitions for the last step of the process, step #4. This task is quite simple: if the starting node of the stage #4 is 4/1 then the optimal (and the only) transition to the last stage is $4/1 \rightarrow 5/1$ with the cost of 13 units. Should we start from node 4/2, the optimal (and the only) transition is $4/2 \rightarrow 5/1$ with the cost of 16 units, and so on. The results of the conditional optimization of the step #4 are tabulated below

Conditional optimization of step 4

Starting node of the stage 4	Final node of the stage 5	Transition costs	Optimal transition	Total cost for this portion of the path
4/1	5/1	13	4/1 → 5/1	13
4/2	5/1	16	4/2 → 5/1	16
4/3	5/1	10	4/3 → 5/1	10
4/4	5/1	11	4/4 → 5/1	11

Let us compile the table representing conditional optimization of the last two steps of the transportation process, namely steps 3 and 4. Assuming that the starting node of the stage 3 is 3/1 then the first available transition within step 3 is $3/1 \rightarrow 4/1$ with the cost of 6 units. At the next step, this transition will be followed by $4/1 \rightarrow 5/1$ and the total cost of both transitions, $3/1 \rightarrow 4/1 \rightarrow 5/1$, is 19 units. Then, consider the second available transition within step 3, $3/1 \rightarrow 4/2$. It comes with the cost of 3 units and must be followed by the transition $4/2 \rightarrow 5/1$ with the total cost of transition $3/1 \rightarrow 4/2 \rightarrow 5/1$ of 19 units. Upon consideration of transitions $3/1 \rightarrow 4/3 \rightarrow 5/1$ and $3/1 \rightarrow 4/4 \rightarrow 5/1$ it could be seen that for 3/1 as the entry point to step 3 the best transition is $3/1 \rightarrow 4/3 \rightarrow 5/1$ with the lowest total cost of 14 units.

Conditional optimization of steps 3 and 4

Starting point of steps 3	Alternative transitions to states	Transition costs	Possible transition	Total cost for two stages	Optimal transition
3/1	4/1	6	3/1 → 4/1 → 5/1	6 + 13 = 19	
	4/2	3	3/1 → 4/2 → 5/1	3 + 16 = 19	
	4/3	4	3/1 → 4/3 → 5/1	4 + 10 = 14	3/1 → 4/3 → 5/1
	4/4	10	3/1 → 4/4 → 5/1	10 + 11 = 21	
3/2	4/1	5	3/2 → 4/1 → 5/1	5 + 13 = 18	
	4/2	6	3/2 → 4/2 → 5/1	6 + 16 = 22	
	4/3	7	3/2 → 4/3 → 5/1	7 + 10 = 17	
	4/4	3	3/2 → 4/4 → 5/1	3 + 11 = 14	3/2 → 4/4 → 5/1
3/3	4/1	11	3/3 → 4/1 → 5/1	11 + 13 = 24	
	4/2	2	3/3 → 4/2 → 5/1	2 + 16 = 18	
	4/3	6	3/3 → 4/3 → 5/1	6 + 10 = 16	3/3 → 4/3 → 5/1
	4/4	8	3/3 → 4/4 → 5/1	8 + 11 = 19	

Now let us compile the table representing conditional optimization of the last three steps of the transportation process, namely step 2 followed by steps 3 and 4. Assume that the starting point of stage 2 is 2/1 and the first available transition is $2/1 \rightarrow 3/1$ with the cost of 8 units. The optimal transition from 3/1 to the last stage has been already established: $3/1 \rightarrow 4/3 \rightarrow 5/1$ and its cost is 14 units, therefore the cost of transition $2/1 \rightarrow 3/1 \rightarrow 4/3 \rightarrow 5/1$ is $8 + 14 = 22$ units. Assume that the starting point is 2/1 and the chosen transition is $2/1 \rightarrow 3/2$ with the cost of 4 units. The already established optimal transition from 3/2 to the last stage is $3/2 \to 4/4 \to 5/1$ with the cost of 14 units, therefore the cost of transition $2/1 \to 3/2 \to 4/4 \to 5/1$ is $4 + 14 = 18$ units. Now assume that the starting point is still 2/1 and the chosen transition is $2/1 \to 3/3$ with the cost of 3 units. The already established optimal transition from 3/3 to the last stage is $3/3 \to 4/3 \to 5/1$ with the cost of 16 units and the total cost is $3 + 16 = 19$ units. This indicates that the optimal path from point 2/1 to 5/1 is $2/1 \to 3/2 \to 4/4 \to 5/1$ with the cost of 18 units. In the similar fashion optimal paths from points 2/2, 2/3 and 2/4 to point 5/1 are to be established. They are: $2/2 \to 3/2 \to 4/4 \to 5/1$ with the cost of 18 units, $2/3 \to 3/2 \to 4/4 \to 5/1$ with the cost of 19 units, and $2/4 \to 3/2 \to 4/4 \to 5/1$ with the cost of 18 units.

Conditional optimization of steps 2, 3 and 4

Starting point of step 2	Alternative transitions to states	Transition costs	Possible transition	Total cost for two stages	Optimal transition
2/1	3/1	8	$2/1 \to 3/1$	$8 + 14 = 22$	
	3/2	4	$2/1 \to 3/2$	$4 + 14 = 18$	$2/1 \to 3/2 \to 4/4 \to 5/1$
	3/3	3	$2/1 \to 3/3$	$3 + 16 = 19$	
2/2	3/1	4	$2/2 \to 3/1$	$4 + 14 = 18$	$2/2 \to 3/1 \to 4/3 \to 5/1$
	3/2	6	$2/2 \to 3/2$	$6 + 14 = 20$	
	3/3	7	$2/2 \to 3/3$	$7 + 16 = 23$	
2/3	3/1	5	$2/3 \to 3/1$	$5 + 14 = 19$	$2/3 \to 3/1 \to 4/3 \to 5/1$
	3/2	6	$2/3 \to 3/2$	$6 + 14 = 20$	
	3/3	8	$2/3 \to 3/3$	$8 + 16 = 24$	
2/4	3/1	9	$2/4 \to 3/1$	$9 + 14 = 23$	
	3/2	5	$2/4 \to 3/2$	$5 + 14 = 19$	$2/4 \to 3/2 \to 4/4 \to 5/1$
	3/3	6	$2/4 \to 3/3$	$6 + 16 = 22$	

Finally, let us compile the table representing optimization of all four steps of the transportation process. Note that the optimization results are not conditional anymore: the transition process is originated at the very particular point, 1/1. Assume that the first available transition is $1/1 \to 2/1$ with the cost of 5 units. The optimal transition from 2/1 to the last stage has been already established: $2/1 \to 3/2 \to 4/4 \to 5/1$ and its cost is 18 units, therefore the cost of transition $1/1 \to 2/1 \to 3/1 \to 4/3 \to 5/1$ is $5 + 18 = 23$ units. Assume that the chosen transition is $1/1 \to 2/2$ with the cost of 3 units. The already established optimal transition from 2/2 to the last stage is $2/2 \to 3/1 \to 4/3 \to 5/1$ with the cost of 18 units, therefore the cost of transition $1/1 \to 2/2 \to 3/1 \to 4/3 \to 5/1$ is $3 + 18 = 21$ units. Now assume that the chosen transition is $1/1 \to 2/3$ with the cost of 1 units. The already established optimal transition from 2/3 to the last stage is $2/3 \to 3/1 \to 4/3 \to 5/1$ with the cost of 19 units and the total cost of transition $1/1 \to 2/3 \to 3/1 \to 4/3 \to 5/1$ is $1 + 19 = 20$ units. Should the chosen transition be $1/1 \to 2/4$ with the cost of 2 units, and since the already established optimal transition from 2/4 to the last stage is $2/4 \to 3/$

$2 \rightarrow 4/4 \rightarrow 5/1$ with the cost of 19 units, the total cost of transition $1/1 \rightarrow 2/4 \rightarrow 3/2 \rightarrow 4/4 \rightarrow 5/1$ is $5 + 19 = 21$ units. This clearly indicates that the optimal path from point $1/1$ to $5/1$ is $1/1 \rightarrow 2/3 \rightarrow 3/1 \rightarrow 4/3 \rightarrow 5/1$. See this analysis summarized in the table below.

Optimization of steps 1, 2, 3 and 4

Starting point of step 1	Alternative transitions to states	Transition costs	Possible transition	Total cost for two stages	Optimal transition
1/1	2/1	5	$1/1 \rightarrow 2/1$	$5 + 18 = 23$	
	2/2	3	$1/1 \rightarrow 2/2$	$3 + 18 = 21$	
	2/3	1	$1/1 \rightarrow 2/3$	$1 + 19 = 20$	$1/1 \rightarrow 2/3 \rightarrow 3/1 \rightarrow$ $4/3 \rightarrow 5/1$
	2/4	2	$1/1 \rightarrow 2/4$	$2 + 19 = 21$	

Consider another quite practical example that ideally lends itself to the application of dynamic programming. It is the optimization of a sequence of manufacturing processes that could be found in chemistry and metallurgy. Each process has its own mathematical description representing quality/quantity of its end product and manufacturing costs as functions of the characteristics of the raw material x_i and control efforts u_i. Consider the mathematical model of i-th manufacturing process within a sequence consisting of N processes:

characteristic of the end material $y_i = y_i(x_i, u_i)$, $i = 1, 2, \ldots, N$

manufacturing cost $q_i = q_i(x_i, u_i)$, $i = 1, 2, \ldots, N$

quality/quantity requirements $y_i^{MIN} \leq y_i \leq y_i^{MAX}$, $i = 1, 2, \ldots, N$

connection to neighboring processes $y_i = x_{i+1}$, $i = 1, 2, \ldots, N$

For simplicity, let us assume that the above functions are scalar and are represented on the basis of their mathematical model by numerical values of y_i and u_i for discretized $x_i = k \cdot \Delta x_i$ and $u_i = m \cdot \Delta u_i$, i.e. $y_i(k, m) = y_i(k \cdot \Delta x_i, m \cdot \Delta u_i)$ and $q_i(k, m) = q_i(k \cdot \Delta x_i, m \cdot \Delta u_i)$ where k, $m = 1, 2, 3, \ldots$. Does this representation of the manufacturing process result in the loss of accuracy? No, providing that the discretization steps Δx_i, Δu_i are judiciously chosen.

Example 4.8 Apply dynamic programming to optimize the operation of a sequence of three manufacturing processes represented by the tabulated description below. Note that the inputs of the individual processes are defined in % assuming that the 100 % value of the respective input corresponds to the maximum value of the output of the previous process. To simplify the problem further, the control efforts are defined not by real numbers, but as "control options." The overall cost of manufacturing is defined as the sum of costs of individual processes. Finally, it could be seen that the specified acceptability limits of the process outputs are different from their feasibility limits that could be seen in the tables.

PROCESS #1					
Output $y_1(x,u)$, $30 \geq y_1 \geq 10$			Cost $q_1(x,u)$		
	U=			U=	
X%	1	2	3	1	2	3
$33 \geq x \geq 0 \rightarrow$	2.000	50.000	18.000	21.000	16.000	73.000
$66 \geq x \geq 34 \rightarrow$	9.000	13.000	19.000	27.000	19.000	130.000
$100 \geq x \geq 67 \rightarrow$	11.000	19.000	31.000	29.000	18.000	21.000

PROCESS #2					
Output $y_2(x,u)$, $70 \geq y_2 \geq 20$			Cost $q_2(x,u)$		
	U=			U=	
X%	1	2	3	1	2	3
$33 \geq x \geq 0 \rightarrow$	8.000	11.000	21.000	70.000	76.000	100.000
$66 \geq x \geq 34 \rightarrow$	24.000	13.000	90.000	61.000	64.000	92.000
$100 \geq x \geq 67 \rightarrow$	1.000	15.000	35.000	55.000	77.000	88.000

PROCESS #3					
Output $y_3(x,u)$, $30 \geq y_3 \geq 15$			Cost $q_3(x,u)$		
	U=			U=	
X%	1	2	3	1	2	3
$33 \geq x \geq 0 \rightarrow$	12.000	50.000	18.000	21.000	16.000	73.000
$66 \geq x \geq 34 \rightarrow$	9.000	13.000	19.000	27.000	29.000	130.000
$100 \geq x \geq 67 \rightarrow$	16.000	19.000	31.000	29.000	28.000	21.000

First, let us address the issue of acceptability limits of the process outputs. Computationally, it could be done by replacing associate cost values by penalties (10^{15}) in the situations when output values are not acceptable—this will automatically exclude some cases from consideration, see the modified tables below

PROCESS #1					
Output $y_1(x,u)$, $30 \geq y_1 \geq 10$				Cost $q_1(x,u)$	
	U=			U=	
X%	1	2	3	1	2	3
$33 \geq x \geq 0 \rightarrow$	2.000	50.000	18.000	.10E+16	.10E+16	73.000
$66 \geq x \geq 34 \rightarrow$	9.000	13.000	19.000	.10E+16	19.000	130.000
$100 \geq x \geq 67 \rightarrow$	11.000	19.000	31.000	29.000	29.000	.10E+16

PROCESS #2					
Output $y_2(x,u)$, $70 \geq y_2 \geq 20$				Cost $q_2(x,u)$	
U=				U=	
X%	1	2	3	1	2	3
$33 \geq x \geq 0 \rightarrow$	8.000	11.000	21.000	.10E+16	.10E+16	100.000
$66 \geq x \geq 34 \rightarrow$	24.000	13.000	90.000	61.000	.10E+16	.10E+16
$100 \geq x \geq 67 \rightarrow$	1.000	15.000	35.000	.10E+16	.10E+16	88.000

PROCESS #3					
Output $y_3(x,u)$, $30 \geq y_3 \geq 15$				Cost $q_3(x,u)$	
U=				U=	
X%	1	2	3	1	2	3
$33 \geq x \geq 0 \rightarrow$	12.000	50.000	18.000	.10E+16	.10E+16	73.000
$66 \geq x \geq 34 \rightarrow$	9.000	13.000	19.000	.10E+16	.10E+16	130.000
$100 \geq x \geq 67 \rightarrow$	16.000	19.000	31.000	29.000	28.000	.10E+16

The following analysis of the problem solution is based on the printout of a specially written computer code. According to the Principle of Optimality, the solution starts from the conditional optimization of the last, third, process. It will provide an optimal recipe for the process operation for every possible grade of the process input. The printout below considers application of various control options when the input of the process is between 0 and 33 % of its maximum attainable value (grade 1). It could be seen that the acceptable value of the process output is obtained only when control option #3 is applied. This defines option #3 as the conditional optimal control option, and the associated cost of 73 units as the conditionally minimal cost.

```
         PROCESS # 3
INP # 1 CONTR# 1 Q= .10000E+16 Y= 12.00
INP # 1 CONTR# 2 Q= .10000E+16 Y= 50.00
INP # 1 CONTR# 3 Q= .73000E+02 Y= 18.00
    OPT: INP # 1, CONTR# 3, QSUM= .73000E+02, Y= 18.00
```

The following printout presents similar results for the situations when the input grade is 2 and 3.

```
INP # 2 CONTR# 1 Q= .10000E+16 Y= 9.00
 INP # 2 CONTR# 2 Q= .10000E+16 Y= 13.00
 INP # 2 CONTR# 3 Q= .13000E+03 Y= 19.00
       OPT: INP # 2, CONTR# 3 QSUM= .13000E+03, Y= 19.00
 INP # 3 CONTR# 1 Q= .29000E+02 Y= 16.00
 INP # 3 CONTR# 2 Q= .28000E+02 Y= 19.00
 INP # 3 CONTR# 3 Q= .10000E+16 Y= 31.00
   OPT: INP # 3, CONTR# 2 QSUM= .28000E+02, Y= 19.00
```

Now consider conditional optimization of process #2. Note that QSUM represents the sum of costs associated with the chosen input and control option of process #2 and the consequent conditionally optimal input + control option of process #3. Consider the application of various control options when the input of process #2 is of grade 1 (i.e. between 0 and 33 % of its maximum attainable value). The resultant QSUM value includes the specific cost at the process #2 and the consequent already known optimal cost at process #3. Since the first two control options are penalized for resulting in unacceptable values of the output, the optimal result is offered by the control option #3 and the accumulated cost value is QSUM $= 73 + 100 = 173$ units. Some additional information seen in the printout addresses the following issue. Note that the action at step #2 has resulted in $y_2 = 21$ units, then how does one determine the consequent action at process #3? It could be seen that the highest y_2 value in the output of process #2 is 90 units. Therefore the output value $y_2 = 21$ falls within 0–33 % of the y_2 range, i.e. $y_2 = 21$ constitutes grade #1 of the input product for process #3. Based on the conditional optimization of process #3, for the input grade #1 control option #1 with the associate cost of 73 units is optimal (see Y $= 21.00 => 1 + (.73000E + 02)$ QSUM $= .17300E + 03$)

```
                       PROCESS # 2
INP  #  1  CONTR#  1  Q = .10000E + 16  Y=  8.00  => 1 + (.73000E + 02)
QSUM = .10000E + 16
INP  #  1  CONTR#  2  Q = .10000E + 16  Y=  11.00  => 1 + (.73000E + 02)
QSUM = .10000E + 16
INP  #  1  CONTR#  3  Q = .10000E + 03  Y=  21.00  => 1 + (.73000E + 02)
QSUM = .17300E + 03
          OPT: INP # 1, CONTR# 3, QSUM = .17300E + 03, Y= 21.00 ==> 1
```

Similar analysis is conducted to perform conditional optimization of process #2 for two other grades of the input.

```
INP  #  2  CONTR#  1  Q = .61000E + 02  Y=  24.00  => 1 + (.73000E + 02)
QSUM = .13400E + 03
INP  #  2  CONTR#  2  Q = .10000E + 16  Y=  13.00  => 1 + (.73000E + 02)
QSUM = .10000E + 16
INP  #  2  CONTR#  3  Q = .10000E + 16  Y=  90.00  => 3 + (.28000E + 02)
QSUM = .10000E + 16
   OPT: INP # 2, CONTR# 1, QSUM = .13400E + 03, Y= 24.00 ==> 1
INP  #  3  CONTR#  1  Q = .10000E + 16  Y=  1.00  => 1 + (.73000E + 02)
QSUM = .10000E + 16
INP  #  3  CONTR#  2  Q = .10000E + 16  Y=  15.00  => 1 + (.73000E + 02)
QSUM = .10000E + 16
INP  #  3  CONTR#  3  Q = .88000E + 02  Y=  35.00  => 2 + (.13000E + 03)
QSUM = .21800E + 03
   OPT: INP # 3, CONTR# 3, QSUM = .21800E + 03, Y= 35.00 ==> 2
```

Consider conditional optimization of process #1, that results in the optimization of the entire combination of three sequential processes. Consider the application of various control options when the input of process #1 is of grade 2 (i.e. between 34 and 66 % of its maximum attainable value). The resultant QSUM value includes the specific cost at the process #1 and the consequent already known optimal costs at process #2 and #3. The first control option results in the unacceptable value of the output and is penalized. The application of control option #2 results in $y_1 = 13$ or #1 grade of the input for process #2, and the cost of 19 units. The already established optimal decisions for this input grade for process #2 come with the cost of 173 units. Consequently QSUM = $19 + 173 = 192$ units. The application of control option #3 results in $y_1 = 19$ (or #2 grade of the input for process #2), and the cost of 130 units. The already established optimal decisions for this input grade for process #2 comes with the cost of 134 units. Therefore QSUM = $130 + 134 = 264$ units. It is clear that the control option #2 is optimal grade #2 of the input material.

```
          PROCESS # 1
INP # 2 CONTR#  1   Q=.10000E+16   Y=     9.00  =>1+(.17300E+03)
QSUM=.10000E+16
INP # 2 CONTR#  2   Q=.19000E+02   Y=    13.00  =>1+(.17300E+03)
QSUM=.19200E+03
INP # 2 CONTR#  3   Q=.13000E+03   Y=    19.00  =>2+(.13400E+03)
QSUM=.26400E+03
          OPT: INP # 2, CONTR# 2, QSUM=.19200E+03, Y=  13.00 ==>1
```

Consider conditional optimization of process #1 when the input of process #1 is of grade #1 and grade #3 is featured below.

```
INP # 1 CONTR#  1   Q=.10000E+16   Y=     2.00  =>1+(.17300E+03)
QSUM=.10000E+16
INP # 1 CONTR#  2   Q=.10000E+16   Y=    50.00  =>3+(.21800E+03)
QSUM=.10000E+16
 INP # 1 CONTR#  3   Q=.73000E+02   Y=    18.00  =>2+(.13400E+03)
QSUM=.20700E+03
   OPT: INP # 1, CONTR# 3, QSUM=.20700E+03, Y=  18.00 ==>2
INP # 3 CONTR#  1   Q=.29000E+02   Y=    11.00  =>1+(.17300E+03)
QSUM=.20200E+03
INP # 3 CONTR#  2   Q=.18000E+02   Y=    19.00  =>2+(.13400E+03)
QSUM=.15200E+03
INP # 3 CONTR#  3   Q=.10000E+16   Y=    31.00  =>2+(.13400E+03)
QSUM=.10000E+16
   OPT: INP # 3, CONTR# 2, QSUM=.15200E+03, Y=  19.00 ==>2
```

Finally, the following printout summarizes the results of the optimization of the entire sequence of three processes for every grade of the raw material.

```
    OPTIMAL PROCESS OPERATION
RAW MATERIAL GRADE:      1      2      3
   PROCESS # 1
CONTROL OPTION:          3      2      2
  OUTPUT =            18.00  13.00  19.00
   PROCESS # 2
CONTROL OPTION:          3      1      3
  OUTPUT =            21.00  24.00  35.00
   PROCESS # 3
CONTROL OPTION:          3      3      2
  OUTPUT =            18.00  19.00  19.00
==================================
 TOTAL COST:         207.00 192.00 152.00
```

4.5.1 Exercise 4.3

Problem 1 Apply dynamic programming to optimize the following sequence of manufacturing processes.

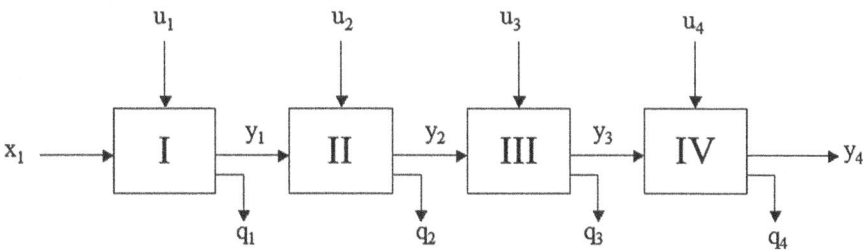

The characteristics of each process are given below:

	$y_1(x,u)$			$q_1(x,u)$			$y_2(x,u)$			$q_2(x,u)$		
	$u=1$	$u=2$	$u=3$	$u=1$	$u=2$	$u=3$	$u=1$	$u=2$	$u=3$	$u=1$	$u=2$	$u=3$
$10 < x \le 40$	25	45	55	25	28	25	65	44	74	13	21	33
$40 < x \le 70$	37	48	63	27	33	27	66	50	81	15	22	37
$70 < x \le 100$	45	58	79	22	24	25	78	62	96	18	28	40

	$y_3(x,u)$			$q_3(x,u)$			$y_4(x,u)$			$q_4(x,u)$		
	$u=1$	$u=2$	$u=3$	$u=1$	$u=2$	$u=3$	$u=1$	$u=2$	$u=3$	$u=1$	$u=2$	$u=3$
$10 < x \le 40$	13	45	92	16	18	9	56	85	97	2	4	3
$40 < x \le 70$	48	18	68	13	17	8	42	61	81	3	6	4
$70 < x \le 100$	81	66	21	10	14	6	21	39	70	4	5	3

It is known that $x_1 = 37$ (units) and the end product must be such that $70 \le y_4 \le 85$. Obtain the optimal choice of control options for each process that would minimize the sum of "local" criteria, $Q = q_1 + q_2 + q_3 + q_4$, and define the corresponding values of the characteristics of the intermediate products.

Problem 2 Use dynamic programming to solve the optimal routing problem based on the graph below featuring the available transitions and the associated costs.

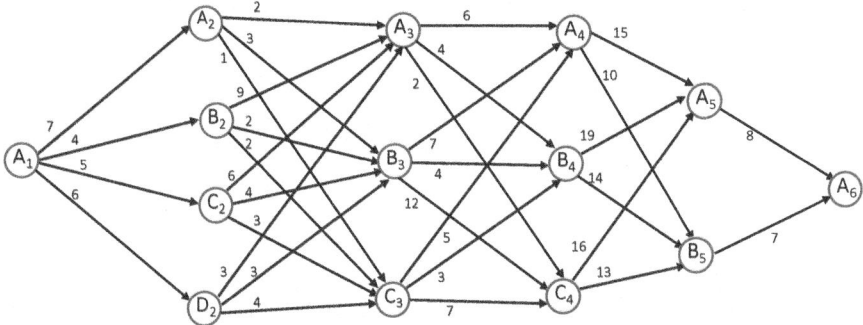

Solutions

Exercise 4.1: Problem 1

The first constraint reflects the requirement on the total weight of the mixture:

$$x_1 + x_2 + x_3 + x_4 + x_5 + x_6 = 500$$

The following expressions represent the required concentrations of each chemical ingredient in the final mixture:

$$Fe = .15 \cdot x_1 + .40 \cdot x_2 + .35 \cdot x_3 + .16 \cdot x_4 + .33 \cdot x_5 + .07 \cdot x_6 \geq .20 \cdot 500$$

$$Zn = .38 \cdot x_1 + .12 \cdot x_2 + .05 \cdot x_3 + .11 \cdot x_4 + .01 \cdot x_5 + .23 \cdot x_6 \geq .10 \cdot 500$$

$$SiO_2 = .41 \cdot x_1 + .40 \cdot x_2 + .27 \cdot x_3 + .21 \cdot x_4 + .60 \cdot x_5 + .45 \cdot x_6 \leq .42 \cdot 500$$

$$Cu = .06 \cdot x_1 + .01 \cdot x_2 + .28 \cdot x_3 + .18 \cdot x_4 + .05 \cdot x_5 + .25 \cdot x_6 \geq .05 \cdot 500$$

The feasibility constraints reflect the availability of the materials. The amount of each material used is equal to the percentage of that material multiplied by the total weight of the end mixture. This must be no greater than the available weight of the material.

$$x_1 \leq 250$$

$$x_2 \leq 590$$

$$x_3 \leq 1000$$

$$x_4 \leq 520$$

$$x_5 \leq 2500$$

$$x_6 \leq 800$$

It should be noted that all variables of this problem are non-negative, but this requirement is very common for linear programming problems and is addressed by the solution algorithm.

The "minimum cost" requirement is addressed as follows:

$$120 \cdot x_1 + 150 \cdot x_2 + 211 \cdot x_3 + 140 \cdot x_4 + 75 \cdot x_5 + 214 \cdot x_6 \rightarrow Min$$

The problem solution is obtained by the use of the linear programming software available in MATLAB. The optimal weight of each material in tons is:

x_1	x_2	x_3	x_4	x_5	x_6	Total	Cost,$
68.2363	0.0000	0.0000	197.5259	234.2378	0.0000	500.0000	53409.80

and the chemical composition of the mixture is

$Fe\%$	$Zn\%$	$SiO_2\%$	$Cu\%$
23.83	10.00	42.00	10.27

Exercise 4.1: Problem 2

For this problem, we were required to use the gradient-based LSM procedure to find the optimal solution of the a coefficients in the following equation.

$$y^{mod}(k) = a_1 \cdot x_1(k) + a_2 \cdot x_2(k) + a_3 \cdot x_3(k) + a_4 \cdot x_4(k)$$

The method for the gradient-based LSM is a simple iterative procedure which "moves" the point representing unknown coefficients in four-dimensional space in the direction toward the minimum value of the criterion. In this case, the criterion, Q, is calculated as the sum of squared values of the discrepancy $e(k) = y(k) - y^{mod}(k)$:

$$A_{new} = \begin{bmatrix} A(1) - \gamma \cdot \dfrac{\Delta Q[A(1), A(2), A(3), A(4)]}{\Delta(1)} \\[1em] A(2) - \gamma \cdot \dfrac{\Delta Q[A(1), A(2), A(3), A(4)]}{\Delta(2)} \\[1em] A(3) - \gamma \cdot \dfrac{\Delta Q[A(1), A(2), A(3), A(4)]}{\Delta(3)} \\[1em] A(4) - \gamma \cdot \dfrac{\Delta Q[A(1), A(2), A(3), A(4)]}{\Delta(4)} \end{bmatrix}$$

where $\dfrac{\Delta Q[A(1), A(2), A(3), A(4)]}{\Delta(i)}$ are estimated partial derivatives of the LSM criterion Q with respect to particular coefficients ($i = 1,2,3,4$) chosen to be 0.0001, and $\gamma > 0$, is a scaling factor. Initially, γ is chosen to be 0.02, however, in the case of an unsuccessful step leading to an increase of criterion Q criterion instead of a decrease, the magnitude of gamma is cut in half. This ensures that the procedure will converge.

The results of this procedure were reached after 250 iterations, starting with zero initial conditions. This procedure could be less accurate but also faster if the termination conditions were made less strict. For this termination condition, the change between the newly generated Q and the previous Q needs to be less than .0000001 in absolute. The optimal result was:

$$A = \begin{bmatrix} 2.0001 \\ 2.9918 \\ -2.0001 \\ 5.0235 \end{bmatrix}$$

Since the coefficient of determination for this model is 0.9996, this is an excellent model of our linear system.

Exercise 4.1: Problem 3

This problem differs from the previous one because of the additional requirement: all model parameters are to be positive. This condition is achieved by the use of penalty functions added to the original LSM criterion Q. In this problem the criterion to be minimized is:

$$Q_1 = Q + \sum_{i=1}^{4} P_i$$

where $Pi = \begin{cases} 0 \ \text{ if } A(i) \geq 0 \\ 10^{10} \cdot A(i)^2 \ \text{ if } A(i) < 0 \end{cases}, i = 1,2,3,4$

The optimal result, with a coefficient of determination of 0.9348, reached after 300 iterations was:

$$A = \begin{bmatrix} 1.0790 \\ 2.4456 \\ .0009 \\ 4.4795 \end{bmatrix}$$

Comparing the coefficient of determination to the one from Problem 2, $0.9348 < 0.9996$ Therefore the coefficients found in Problem 2 are a better representation of the actual system. Since the actual system includes a negative coefficient for a_3, not allowing negative coefficients in Problem 3 impacted the ability of the optimization to get close to the actual coefficient values.

Exercise 4.2: Problem 1

For this problem, we were given the following transfer function of the controlled plant, and were required to use the Simplex code provided to find the optimal coefficients of a PID controller. The PID controller was configured to use the system error as its input for the controller.

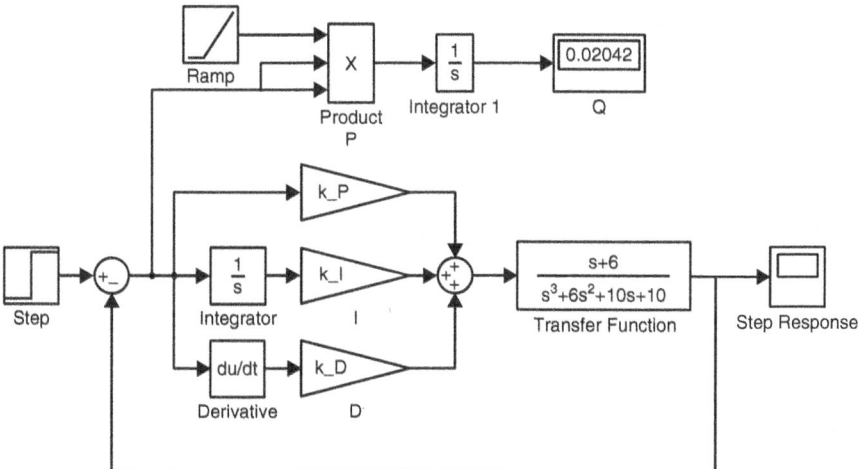

As could be seen, the optimization criterion, known as "integral-time-error-squared" was chosen. The Simplex procedure began with zero initial coefficient values, and progressively changed these values, minimizing the criterion. All 30 iteration of the Simplex procedure are shown below.

Iteration	k_P	k_I	k_D	Q
1	0.000	0.000	0.000	50.00000
2	0.094	0.024	0.024	38.18000
3	0.024	0.094	0.024	25.61000
4	0.024	0.024	0.094	41.24000
5	0.094	0.094	0.094	24.15000
6	0.141	0.141	0.141	17.33000
7	0.149	0.149	0.031	16.29000
8	0.212	0.212	0.000	10.76000

(continued)

Iteration	k_P	k_I	k_D	Q
9	0.157	0.275	0.086	7.86200
10	0.189	0.401	0.118	4.21000
11	0.338	0.409	0.149	3.87000
12	0.495	0.566	0.212	2.06500
13	0.456	0.644	0.079	1.62100
14	0.613	0.896	0.047	0.86220
15	0.652	1.029	0.251	0.72740
16	0.872	1.438	0.377	0.48830
17	1.131	1.532	0.306	0.39070
18	1.603	2.098	0.401	0.29570
19	1.563	2.388	0.338	0.34080
20	2.079	3.054	0.697	0.29080
21	2.813	4.133	1.021	0.26310
22	3.114	4.308	0.796	0.25040
23	4.235	5.743	1.006	0.21720
24	4.203	5.594	1.281	0.19470
25	5.523	7.197	1.752	0.15960
26	6.778	9.284	2.119	0.14560
27	9.365	12.877	2.978	0.11670
28	9.936	13.079	2.802	0.10360
29	13.498	17.552	3.693	0.08105
30	14.689	19.341	4.609	0.08550

The following plots illustrate gradual improvement of the closed-loop step response of the system, iterations 1, 2, 8, 12, 14, 16 are shown below. Technically, the procedure could be terminated after the 16-th iteration when the design requirements were met.

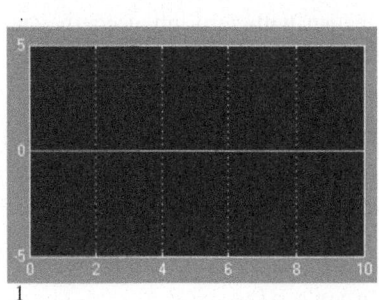

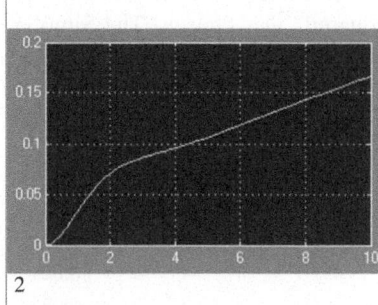

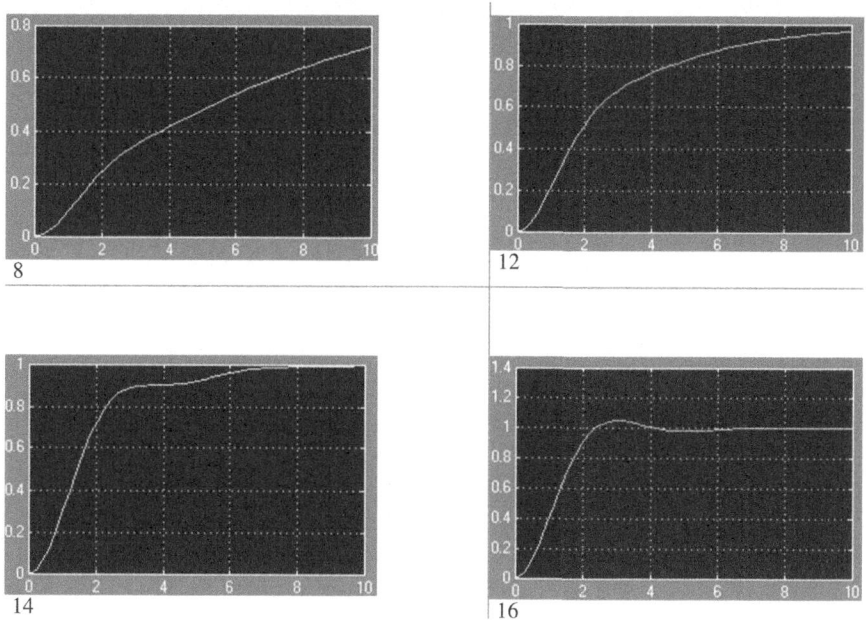

8

12

14

16

Exercise 4.2: Problem 2

The task is to estimate parameters a_1, a_2, a_3, a_4, a_5 of the following model

$$y^{MOD} = a_1 x_1 e^{a_2 x_2} + a_3^{(a_4 x_3 + a_5)}$$

based on the input–output data presented in the table below. It could be seen that while the Least Squares Method (LSM) is to be applied, due to the nonlinearity of the model, traditional LSM equation is unusable and the LSM criterion,

$$Q(a_1,\ a_2,\ a_3,\ a_4,\ a_5) = \sum_i \left[y(i) - y^{MOD}(i) \right]^2$$

could be minimized only by a genetic optimization (GO) procedure.

Assume that a "generation" size is 20. To start the procedure, begin with 20 randomly generated sets of 5 coefficients compiled into in a 20×5 matrix, with each row representing a particular set of coefficients (an individual). In each generation these 20 "individuals" will become the source of off-spring and mutants. Each combination of two individuals within a generation results in 5 off-spring. Each individual in the generation originates 5 mutants.

It could be seen that this process results in an immense number of new individuals, however, each newly created individual is subjected to the "fitness test" and only the 20 most fit individuals are included in the next generation. Each "next generation" is subjected to the same treatment until some termination conditions are met.

Below are several successive generation created by GO and accompanying values of the coefficient of determination of the resultant model.

Initial generation

a_1	a_2	a_3	a_4	a_5
2.8985	1.1513	1.3472	0.7978	−1.7259
1.6345	0.3528	−3.1390	−0.4051	−1.6175
4.0328	−0.2817	−0.8104	−0.2262	2.0505
0.5651	−0.3717	−1.4612	2.1874	0.2249
0.0178	−2.6064	0.7837	2.0933	0.9262
−0.4143	1.3217	0.6731	−1.0293	0.6581
0.3766	2.1826	2.5247	−1.7790	−0.3017
1.9690	−1.3128	−0.5600	2.6983	−2.0503
1.5223	−1.7655	−1.0972	−6.0455	1.9243
−1.9364	2.2881	−2.5293	1.9482	−1.2985
−0.1199	3.7093	0.4981	0.0415	−0.1648
−2.6940	2.6746	−2.0002	1.0056	−0.6448
−1.6309	3.0015	−1.4742	2.2930	2.2985
−1.6146	3.2141	3.1189	1.3045	−4.4747
3.7037	−0.5679	0.2711	−1.9776	0.4899
1.3725	−0.5816	0.3215	−2.4576	1.0954
0.9418	2.1572	−0.3685	3.4800	−3.1552
−2.9340	−4.9244	−0.0846	0.5385	−1.4570
3.5418	−0.5850	−0.3802	−4.0273	0.3840
0.5160	−4.4928	−0.1777	3.0757	−1.5804

Generation 1

a_1	a_2	a_3	a_4	a_5	Determination coefficient
2.8985	1.1513	1.3472	0.7978	−1.7259	0.8825
2.2881	1.2683	1.3268	0.3294	−0.2899	0.8756
3.1209	1.2110	1.8546	−1.8284	0.2756	0.8755
2.5022	1.2605	0.6157	−0.9998	1.3549	0.8656
1.8744	1.2679	−0.9757	0.6641	2.0549	0.8601
1.1644	1.4379	−1.7683	−1.0123	−0.8917	0.8532
1.6140	1.2957	−0.5244	−0.5290	1.4760	0.8408
0.7947	1.6388	0.9696	−2.4417	0.6111	0.8397
2.0885	1.1947	0.8106	−0.2046	−0.5338	0.8384
0.8033	1.5508	2.2729	−0.0392	−0.5913	0.8352

(continued)

a_1	a_2	a_3	a_4	a_5	Determination coefficient
2.5520	1.1144	0.4245	−0.3246	0.5987	0.8319
1.1761	1.5435	1.4602	−2.2510	−0.0922	0.8282
3.4343	1.0007	−0.5871	0.2609	1.1236	0.8243
1.6577	1.2627	−0.9837	1.3665	−0.4513	0.8215
1.3407	1.4972	−2.5164	0.8116	−2.9722	0.8175
2.9522	1.0467	0.1449	0.5600	1.8093	0.8157
0.6689	1.5793	−0.7417	2.8365	0.1993	0.8074
2.7650	1.0530	1.1006	0.2516	−1.6950	0.7990
1.8029	1.2170	−1.9619	1.3318	−1.5114	0.7959
2.8907	1.2643	0.7140	0.1415	−2.4496	0.7925

Generation 3

a_1	a_2	a_3	a_4	a_5	Determination coefficient
3.2330	1.1311	0.1180	0.1429	−1.5847	0.9203
3.4089	1.1194	0.1505	0.2626	−1.3844	0.9178
3.0811	1.1600	0.1540	0.3674	−1.2171	0.9152
3.2224	1.1584	0.1650	0.2263	−1.4428	0.9142
3.2893	1.1546	0.1817	0.8279	−0.1388	0.9137
2.5298	1.2368	0.0907	0.5090	−0.1884	0.9114
3.0611	1.1398	0.0935	0.0080	−1.5563	0.9112
2.4017	1.2431	0.2474	0.7796	−0.6685	0.9102
3.3596	1.1058	0.1555	0.1667	−1.3748	0.9095
2.8410	1.1452	0.1536	0.0956	−1.7017	0.9089
2.4652	1.2277	0.0785	0.3539	−0.4969	0.9087
2.2753	1.2568	0.2190	0.5816	−0.8525	0.9073
2.3878	1.2545	0.1447	0.5450	−0.3459	0.9068
2.6861	1.1952	0.2399	0.2100	−1.6736	0.9057
3.1428	1.1634	0.2227	0.1615	−1.6702	0.9056
2.6985	1.1850	0.2020	0.1200	−1.6083	0.9054
2.4388	1.2495	0.0313	0.3191	−0.1237	0.9054
3.2899	1.1293	0.2615	0.2999	−1.4908	0.9051
2.3510	1.2336	0.2180	0.3131	−1.4478	0.9048
2.8936	1.1421	0.1960	0.1882	−1.6563	0.9044

Generation 5

a_1	a_2	a_3	a_4	a_5	Determination coefficient
3.2187	1.1474	0.1081	0.1070	−1.5610	0.9243
3.2307	1.1444	0.1065	0.0930	−1.5475	0.9242
3.1887	1.1535	0.1066	0.0969	−1.5458	0.9241
3.1930	1.1516	0.1153	0.0942	−1.6730	0.9241
3.1678	1.1503	0.1164	0.1062	−1.5840	0.9239
3.3344	1.1401	0.1163	0.1289	−1.5773	0.9239
3.2312	1.1458	0.1056	0.0839	−1.5436	0.9239

(continued)

a_1	a_2	a_3	a_4	a_5	Determination coefficient
3.1755	1.1476	0.1139	0.0927	−1.5828	0.9238
3.1045	1.1533	0.1032	0.0814	−1.5745	0.9238
3.1020	1.1555	0.1101	0.0826	−1.5823	0.9238
3.1053	1.1523	0.1109	0.0850	−1.5895	0.9237
3.2088	1.1392	0.1033	0.0831	−1.5563	0.9237
3.3173	1.1454	0.1160	0.1322	−1.5786	0.9236
3.1123	1.1533	0.1164	0.1001	−1.5813	0.9236
3.2996	1.1295	0.1009	0.0858	−1.5604	0.9236
3.3068	1.1395	0.1328	0.1453	−1.6407	0.9236
3.3615	1.1333	0.1098	0.1346	−1.5013	0.9235
3.2392	1.1462	0.1213	0.1365	−1.5729	0.9235
3.2177	1.1370	0.1105	0.0952	−1.5843	0.9235
3.0982	1.1542	0.1172	0.1026	−1.5828	0.9235

Generation 7

a_1	a_2	a_3	a_4	a_5	Determination coefficient
3.2964	1.1409	0.1069	0.1003	−1.5657	0.9244
3.2921	1.1406	0.1063	0.0987	−1.5740	0.9244
3.2906	1.1400	0.1057	0.1014	−1.5538	0.9244
3.2617	1.1426	0.1080	0.1004	−1.5746	0.9244
3.2937	1.1406	0.1141	0.1063	−1.6061	0.9244
3.2497	1.1445	0.1063	0.1002	−1.5612	0.9244
3.2837	1.1424	0.1048	0.0966	−1.5579	0.9244
3.2246	1.1478	0.1122	0.1016	−1.6072	0.9244
3.2764	1.1418	0.1104	0.1047	−1.5810	0.9244
3.2292	1.1473	0.1077	0.1007	−1.5728	0.9244
3.2272	1.1467	0.1149	0.1005	−1.6283	0.9244
3.2825	1.1422	0.1085	0.0975	−1.5793	0.9244
3.2831	1.1400	0.1081	0.1029	−1.5699	0.9244
3.2317	1.1464	0.1090	0.1021	−1.5838	0.9244
3.2291	1.1462	0.1074	0.0996	−1.5691	0.9244
3.2202	1.1473	0.1083	0.1002	−1.5767	0.9244
3.2817	1.1425	0.1073	0.1012	−1.5615	0.9244
3.2629	1.1422	0.1116	0.1037	−1.5985	0.9244
3.2778	1.1426	0.1187	0.1053	−1.6354	0.9244
3.2414	1.1462	0.1072	0.1030	−1.5708	0.9244

Generation 9

a_1	a_2	a_3	a_4	a_5	Determination coefficient
3.2931	1.1403	0.1067	0.0996	−1.5671	0.9244
3.2946	1.1402	0.1062	0.0997	−1.5652	0.9244
3.2915	1.1403	0.1077	0.1000	−1.5749	0.9244
3.2941	1.1402	0.1060	0.0996	−1.5644	0.9244
3.2921	1.1404	0.1069	0.0999	−1.5686	0.9244
3.2934	1.1404	0.1059	0.0986	−1.5632	0.9244
3.2905	1.1404	0.1056	0.0991	−1.5608	0.9244
3.2936	1.1404	0.1113	0.1011	−1.5986	0.9244
3.2937	1.1404	0.1064	0.0998	−1.5654	0.9244
3.2936	1.1402	0.1061	0.1000	−1.5628	0.9244
3.2917	1.1402	0.1062	0.0996	−1.5629	0.9244
3.2895	1.1405	0.1055	0.0987	−1.5605	0.9244
3.2943	1.1403	0.1063	0.1001	−1.5638	0.9244
3.2900	1.1404	0.1057	0.0986	−1.5612	0.9244
3.2913	1.1403	0.1060	0.0996	−1.5622	0.9244
3.2918	1.1404	0.1064	0.0999	−1.5648	0.9244
3.2902	1.1402	0.1056	0.0985	−1.5612	0.9244
3.2901	1.1406	0.1072	0.0997	−1.5719	0.9244
3.2918	1.1406	0.1057	0.0988	−1.5631	0.9244
3.2933	1.1405	0.1065	0.0999	−1.5674	0.9244

After 9 iterations of the GO procedure stopped, and the resultant value of the coefficient of determination was 0.9244 for coefficients [3.2931, 1.1403, 0.1067, 0.0996, −1.5671], and the model expression is: $y^{MOD} = 3.2931 x_1 e^{1.1403 x_2} + 0.1067^{0.0996 x_3 - 1.5671}$

Exercise 4.3: Problem 1

Conditional optimization of Process IV:

If $x_4 = [10, 40]$
      Choose $u_4 = 2$
      Cost $= 4$

If $x_4 = (40, 70]$
      Choose $u_4 = 3$
      Cost $= 4$

If $x_4 = (70, 100]$
  Choose $u_4 = 3$
  Cost $= 3$

Conditional optimization of Process III and Process IV:

If $x_3 = [10, 40]$
  If $u_3 = 1$, cost $= 16 + \text{cost}(x_4 = 13) = 16 + 4 = 20$
  If $u_3 = 2$, cost $= 18 + \text{cost}(x_4 = 45) = 18 + 4 = 22$
  If $u_3 = 3$, cost $= 9 + \text{cost}(x_4 = 92) = 9 + 3 = \mathbf{12}$ (optimal)
  Choose $u_3 = 3$
  Cost $= 12$

If $x_3 = (40, 70]$
  If $u_3 = 1$, cost $= 13 + \text{cost}(x_4 = 48) = 13 + 4 = 17$
  If $u_3 = 2$, cost $= 17 + \text{cost}(x_4 = 18) = 17 + 4 = 21$
  If $u_3 = 3$, cost $= 8 + \text{cost}(x_4 = 68) = 8 + 4 = \mathbf{12}$ (optimal)
  Choose $u_3 = 3$
  Cost $= 12$

If $x_3 = (70, 100]$
  If $u_3 = 1$, cost $= 10 + \text{cost}(x_4 = 81) = 10 + 3 = 13$
  If $u_3 = 2$, cost $= 14 + \text{cost}(x_4 = 66) = 14 + 4 = 18$
  If $u_3 = 3$, cost $= 6 + \text{cost}(x_4 = 21) = 6 + 4 = \mathbf{10}$ (optimal)
  Choose $u_3 = 3$
  Cost $= 10$

Conditional optimization of Process II, Process III and Process IV:

If $x_2 = [10, 40]$
  If $u_2 = 1$, cost $= 13 + \text{cost}(x_3 = 65) = 13 + 12 = \mathbf{25}$ (optimal)
  If $u_2 = 2$, cost $= 21 + \text{cost}(x_3 = 44) = 21 + 12 = 33$
  If $u_2 = 3$, cost $= 33 + \text{cost}(x_3 = 74) = 33 + 10 = 43$
  Choose $u_2 = 1$
  Cost $= 25$

If $x_2 = (40, 70]$
  If $u_2 = 1$, cost $= 15 + \text{cost}(x_3 = 66) = 15 + 12 = \mathbf{27}$ (optimal)
  If $u_2 = 2$, cost $= 22 + \text{cost}(x_3 = 50) = 22 + 12 = 33$
  If $u_2 = 3$, cost $= 37 + \text{cost}(x_3 = 81) = 37 + 10 = 47$
  Choose $u_2 = 1$
  Cost $= 27$

If $x_2 = (70, 100]$

    If $u_2 = 1$, cost $= 18 + \text{cost}(x_3 = 78) = 18 + 10 = \mathbf{28}$ (optimal)

    If $u_2 = 2$, cost $= 28 + \text{cost}(x_3 = 62) = 28 + 12 = 40$

    If $u_2 = 3$, cost $= 40 + \text{cost}(x_3 = 96) = 40 + 10 = 50$

    Choose $u_2 = 1$

    Cost $= 28$

Conditional optimization of Process I, Process II, Process III, and Process IV:

If $x_1 = [10, 40]$

    If $u_1 = 1$, cost $= 25 + \text{cost}(x_2 = 25) = 25 + 25 = \mathbf{50}$ (optimal)

    If $u_1 = 2$, cost $= 28 + \text{cost}(x_2 = 45) = 28 + 27 = 55$

    If $u_1 = 3$, cost $= 25 + \text{cost}(x_2 = 55) = 25 + 27 = 52$

    Choose $u_1 = 1$

    PATH: $u_1 = 1 \rightarrow u_2 = 1 \rightarrow u_3 = 3 \rightarrow u_4 = 3$

    Cost $= 50$

If $x_1 = (40, 70]$

    If $u_1 = 1$, cost $= 27 + \text{cost}(x_2 = 37) = 27 + 25 = \mathbf{52}$ (optimal)

    If $u_1 = 2$, cost $= 33 + \text{cost}(x_2 = 48) = 33 + 27 = 60$

    If $u_1 = 3$, cost $= 27 + \text{cost}(x_2 = 63) = 27 + 28 = 55$

    Choose $u_1 = 1$

    PATH: $u_1 = 1 \rightarrow u_2 = 1 \rightarrow u_3 = 3 \rightarrow u_4 = 3$

    Cost $= 52$

If $x_1 = (70, 100]$

    If $u_1 = 1$, cost $= 22 + \text{cost}(x_2 = 45) = 22 + 27 = \mathbf{49}$ (optimal)

    If $u_1 = 2$, cost $= 24 + \text{cost}(x_2 = 58) = 24 + 27 = 51$

    If $u_1 = 3$, cost $= 25 + \text{cost}(x_2 = 79) = 25 + 28 = 53$

    Choose $u_1 = 1$

    PATH: $u_1 = 1 \rightarrow u_2 = 1 \rightarrow u_3 = 3 \rightarrow u_4 = 3$

    Cost $= 49$

Optimal Plan

Since $x = 37$, the optimal path is $u_1 = 1 \rightarrow u_2 = 1 \rightarrow u_3 = 3 \rightarrow u_4 = 3$ and the cost $= 50$.

Exercise 4.3: Problem 2

Conditional optimization of particular stages of the process starting from the last stage:

Stage $5 \rightarrow 6$
If at $A5 \rightarrow A6 - \text{Cost} = 8$ (optimal)
If at $B5 \rightarrow A6 - \text{Cost} = 7$ (optimal)

Stage $4 \rightarrow 5$
If at $A4 \rightarrow \text{A5} - \text{Cost} = 15 + 8 = 23$
$\quad\quad\quad \rightarrow B5 - \text{Cost} = 10 + 7 = 17$ (optimal)
If at $B4 \rightarrow \text{A5} - \text{Cost} = 19 + 8 = 27$
$\quad\quad\quad \rightarrow B5 - \text{Cost} = 14 + 7 = 21$ (optimal)
If at $C4 \rightarrow \text{A5} - \text{Cost} = 16 + 8 = 24$
$\quad\quad\quad \rightarrow B5 - \text{Cost} = 13 + 7 = 20$ (optimal)

Stage $3 \rightarrow 4$
If at $A3 \rightarrow \text{A4} - \text{Cost} = 6 + 17 = 23$
$\quad\quad\quad \rightarrow \text{B4} - \text{Cost} = 4 + 21 = 25$
$\quad\quad\quad \rightarrow C4 - \text{Cost} = 2 + 20 = 22$ (optimal)
If at $B3 \rightarrow A4 - \text{Cost} = 7 + 17 = 24$ (optimal)
$\quad\quad\quad \rightarrow \text{B4} - \text{Cost} = 4 + 21 = 25$
$\quad\quad\quad \rightarrow \text{C4} - \text{Cost} = 12 + 20 = 32$
If at $C3 \rightarrow A4 - \text{Cost} = 5 + 17 = 22$ (optimal)
$\quad\quad\quad \rightarrow B4 - \text{Cost} = 3 + 21 = 24$
$\quad\quad\quad \rightarrow C4 - \text{Cost} = 7 + 20 = 27$

Stage $2 \rightarrow 3$
If at $A2 \rightarrow \text{A3} - \text{Cost} = 2 + 22 = 24$
$\quad\quad\quad \rightarrow \text{B3} - \text{Cost} = 3 + 24 = 27$
$\quad\quad\quad \rightarrow C3 - \text{Cost} = 1 + 22 = 23$ (optimal)
If at $B2 \rightarrow \text{A3} - \text{Cost} = 9 + 22 = 31$
$\quad\quad\quad \rightarrow \text{B3} - \text{Cost} = 2 + 24 = 26$
$\quad\quad\quad \rightarrow C3 - \text{Cost} = 2 + 22 = 24$ (optimal)
If at $C2 \rightarrow \text{A3} - \text{Cost} = 6 + 22 = 28$
$\quad\quad\quad \rightarrow \text{B3} - \text{Cost} = 4 + 24 = 28$
$\quad\quad\quad \rightarrow C3 - \text{Cost} = 3 + 22 = 25$ (optimal)
If at $D2 \rightarrow A3 - \text{Cost} = 3 + 22 = 25$ (optimal)
$\quad\quad\quad \rightarrow \text{B3} - \text{Cost} = 3 + 24 = 27$
$\quad\quad\quad \rightarrow \text{C3} - \text{Cost} = 4 + 22 = 26$

Stage $1 \rightarrow 2$
If at A1 $\rightarrow$ ~~A2 — Cost $= 7 + 23 = 30$~~
$\qquad\quad \rightarrow$ B2 − Cost $= 4 + 24 = 28$ (optimal)
$\qquad\quad \rightarrow$ ~~C2 — Cost $= 5 + 25 = 30$~~
$\qquad\quad \rightarrow$ ~~D2 — Cost $= 6 + 25 = 31$~~

Optimal Path

A1 $\rightarrow$ B2 $\rightarrow$ C3 $\rightarrow$ A4 $\rightarrow$ B5 $\rightarrow$ A6, Cost: 28

Bibliography

http://www.onlinecalculatorfree.org/linear-programming-solver.html
MATLAB: https://www.mathworks.com/products/matlab-home/
Simulink: http://www.mathworks.com/products/simulink/
VISSIM: http://www.vissim.com/

Index

© Springer International Publishing Switzerland 2016   251
V.A. Skormin, *Introduction to Process Control*, Springer Texts
in Business and Economics, DOI 10.1007/978-3-319-42258-9

The manufacturer's authorised representative in the EU is Springer
Nature Customer Service Centre GmbH, Europaplatz 3, 69115 Heidelberg,
Germany. If you have any concerns regarding our products, please
contact ProductSafety@springernature.com

Printed and bound by CPI Group (UK) Ltd, Croydon, CR0 4YY

23/04/2026
02095639-0001